PACEMAKER'S®

PASSAGE TO Basic Math

Globe Fearon Educational Publisher
A Division of Simon & Schuster
Upper Saddle River, New Jersey

Director Editorial & Marketing, Special Education: Diane Galen
Market Manager: Susan McLaughlin
Executive Editor: Joan Carrafiello
Project Editor: Stephanie Petron Cahill
Contributing Editors: Jennifer McCarthy, Elena Petron
Editorial Development: Lake Publishing, Inc.
Editorial Assistants: Derrell Bradford, Ryan Jones
Production Director: Kurt Scherwatzky
Production Editor: John Roberts
Art Direction and Cover Design: Pat Smythe, Armondo Baez
Page Design: Margarita Giammanco
Electronic Page Production: Burmar Technical Corporation

Teacher Reviewers:

Beth Bina Carlson,
Special Needs Teacher,
Central High,
Grand Forks, North Dakota

Carolyn Lambert,
Special Education Teacher,
Lower Pioneer Valley Educational Collaborative,
Wilbraham, Massachusetts

Dorie L. Knaub, BS, MA.,
Consultant,
Downey Unified School District,
Downey, California

Printed in the United States of America
5 6 7 8 9 10 00 99

ISBN: 0-8359-3464-0

Globe Fearon Educational Publisher
A Division of Simon & Schuster
Upper Saddle River, New Jersey

Contents

Note to the Student

Math is used everywhere in the world around us—in making change, in comparing prices, in measuring for a recipe, in finding how far it is to a park or even to a planet. These are only a few of the ways we use math.

As you work through *Passage to Basic Math* you will learn the number skills you need to add, subtract, multiply, and divide. You will work with percents and solve many problems. All of these important skills are taught in small parts that are easy to understand. When you finish, you will have the tools it takes to succeed at math in everyday life.

Each lesson of *Passage to Basic Math* starts with Words to Know. This section shows you words you may not have seen before. Then, when you see them in the lesson, you will have some idea of what they mean.

The lessons have small parts. They are named by letters. Each part teaches you something new through clear examples. Every part also contains exercises. Exercises ask you to use what you have just learned. As you do them, you can show your work and answers on a separate sheet of paper.

There are other study aids in the lessons, too. Reminders point out things you might need to follow the examples. Other notes in the margin give you helpful hints.

We wish you well on your *Passage to Basic Math*. Our success comes from your success.

Unit 1 Whole Numbers

Lesson 1 Using a Number Line

Words to Know

whole numbers 0, 1, 2, 3, 4, 5, 6, 7, and so on
number line numbers shown in order as points on a line
even number a number that ends in 0, 2, 4, 6, or 8
odd number a number that ends in 1, 3, 5, 7, or 9

Have you ever used a ruler or a measuring tape? If so, you have used a number line. A number line is useful. It can help you count, add, or subtract.

A. The numbers 0, 1, 2, 3, 4, 5, 6, and so on are called **whole numbers.** Whole numbers are used for counting. Whole numbers tell how many or how much. You can read "how many or how much" in numbers or words.

We won **nine** games this season.
I ran **9** miles.
There are **eight** students in the library.
The ladder is **8** feet long.

Reminder

Write a new sentence on a separate sheet of paper.

Read each sentence. Find "how many or how much." Rewrite the word as a number.

1. The rug is six feet long.
2. My school is five miles away.
3. I spent two dollars on trading cards.
4. She measured four ounces of milk.
5. That building is seven stories tall.
6. We played tennis for one hour.

Read each sentence. Find "how many or how much." Rewrite the number as a word.

7. She ate 2 muffins.
8. I saw 6 birds by the window.
9. Henry plays 3 different sports.
10. Lucy saw that movie 7 times.
11. They ordered 8 CDs from the music club.
12. There were 5 different desserts on the menu.

B.

A **number line** shows numbers as points on a line. The numbers are in order from 0 to 10 or higher. Look at the number line. The number 1 is smaller than 10. The number 10 is larger than 5. The numbers get larger as you go along the line.

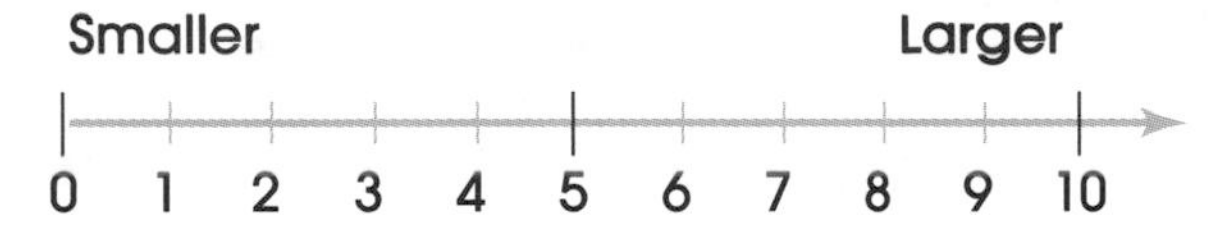

Look at the number line below. What are the missing numbers? Draw a number line. Write in all the numbers.

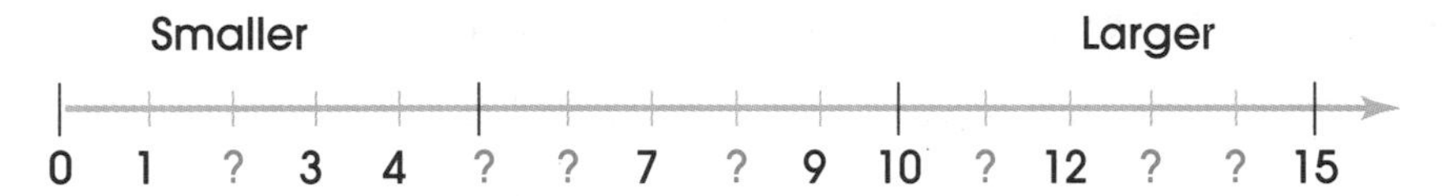

C.

There are two kinds of numbers, even numbers and odd numbers. **Even numbers** end in 0, 2, 4, 6, and 8. Look at the even numbers on the number line below.

0 1 2 3 4 5 6 7 8 9 10

Look at the chart below. List all the even numbers.

1	2	3	4	5	6	7	8	9	10
11	12	13	14	15	16	17	18	19	20
21	22	23	24	25	26	27	28	29	30
31	32	33	34	35	36	37	38	39	40

D.

Odd numbers end in 1, 3, 5, 7, and 9. Look at the odd numbers on the number line below.

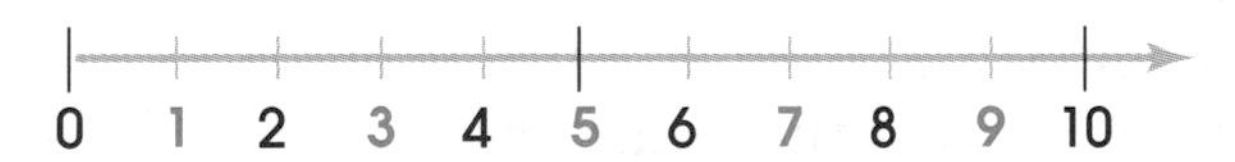

Look at the chart below. List all the odd numbers.

41	42	43	44	45	46	47	48	49	50
51	52	53	54	55	56	57	58	59	60
61	62	63	64	65	66	67	68	69	70
71	72	73	74	75	76	77	78	79	80

Lesson 2 Place Value

Words to Know

digit used to write a number: 0, 1, 2, 3, 4, 5, 6, 7, 8, 9
place value value or amount of a digit; a digit's worth
rename to write a number in a different way

The word *digit* sounds like DIJ-uht.

A **digit** is used to write a number. The ten digits are 0, 1, 2, 3, 4, 5, 6, 7, 8, 9. These digits can be used to write any number.

A. On its own, a digit stands for an amount. Look at the example below. The digit 4 stands for more stars than the 3.

★★★ ⟶ 3   ★★★★ ⟶ 4

Most numbers are made up of two or more digits. A digit's place in a number tells its value or how much it is worth. This is called **place value.** A digit's place helps you know how to say a number. Look at the number 83.

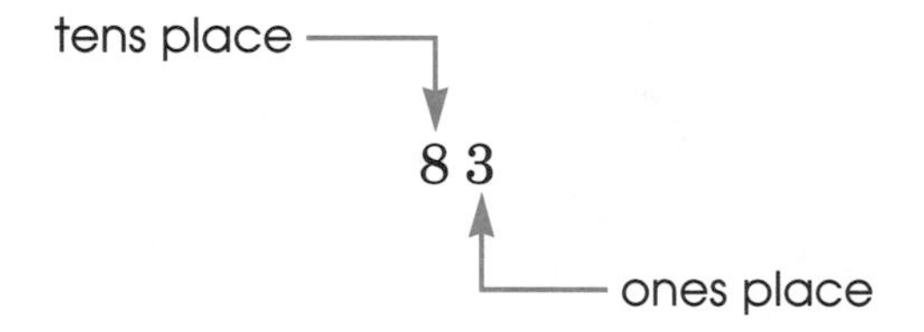

The 8 is in the tens place. It is worth 8 tens or eighty. The three is in the ones place. It is worth 3 ones or three. To say 83, you say "eighty-three."

A place value chart can help you find the value of larger numbers. Look at the numbers in the chart in the margin. Each number is made up of the same three digits. Yet each number is different.

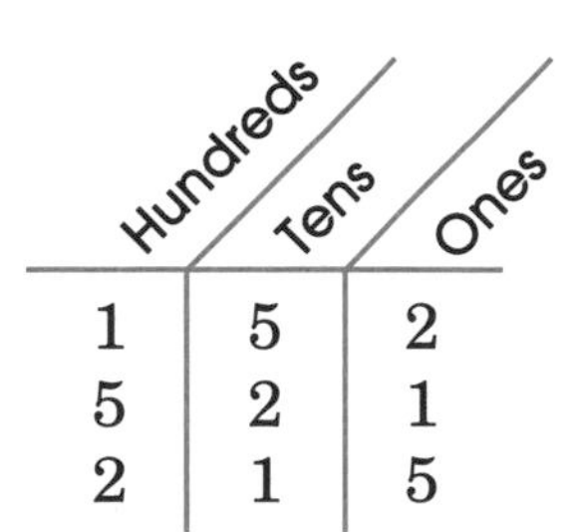

Hundreds	Tens	Ones
1	5	2
5	2	1
2	1	5

What makes the numbers different? The digits are in different places. Notice that each digit falls in a certain place. Look at the number 152. It shows 1 hundred and 5 tens and 2 ones. You read 152 as "one hundred fifty-two."

Read each number. Find the digit 5. Write its place in each number.

1. In 152, the 5 is in the _____ place.
2. In 521, the 5 is in the _____ place.
3. In 215, the 5 is in the _____ place.

B. You have seen that digits can be in different places: a ones place, a tens place, and a hundreds place. Knowing the place values helps you say the number.

Read the number 483. Answer the questions.

1. Which digit is in the hundreds place?
2. Which digit is in the tens place?
3. Which digit is in the ones place?
4. How do you say the number?

C. You can **rename** a number. That means to show it in a different way.

56 can be renamed as 5 tens + 6 ones.
792 can be renamed as 7 hundreds + 9 tens + 2 ones.

Read each renamed number. What number is being renamed? Write your answer in a complete sentence.

1. 3 tens + 4 ones
2. 6 tens + 2 ones
3. 5 hundreds + 4 tens + 6 ones
4. 3 hundreds + 9 tens + 8 ones

D. Larger numbers have more digits and more places. Look at the number 3,210 on this place value chart.

As numbers get bigger, they use more places: thousands, ten thousands, hundred thousands, and even millions.

Hundred thousands	Ten thousands	Thousands	Hundreds	Tens	Ones
		3,	2	1	0
		5,	6	2	3

The comma in a number tells you where the thousands place starts. Look at each digit's place value in the chart above. How do you say this number?

3,210 ⟶ three thousand two hundred ten

Use the chart above to answer the questions about 5,623.

1. Which digit is in the thousands place?
2. Which digit is in the hundreds place?
3. Which digit is in the tens place?
4. Which digit is in the ones place?
5. How do you say 5,623 in words?

Lesson 3 Rounding Numbers

Words to Know

rounding numbers changing numbers to tell *about* how many in tens, hundreds, or thousands

Sometimes numbers need to be exact. Sometimes they don't. You may need to make numbers easy to work with. If so, you might say *about* how many.

A.

Rounding numbers lets you tell *about* instead of *exactly* how many. When you round a number, you can change it to show about how many tens, hundreds, or thousands. Look at the example below.

Randy has exactly **37** music CDs.
"I have about **40** CDs," Randy said.
He rounded up to the nearest **ten**.

You can use a number line to round numbers to the nearest ten.

- 52 is closer to 50 than to 60.
- 52 rounds down to **50**.

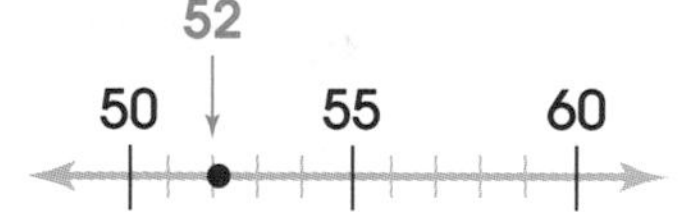

Rounding down makes a smaller number. 50 is less than 52.

- 55 is exactly halfway between 50 and 60.
- 55 rounds up to **60**.

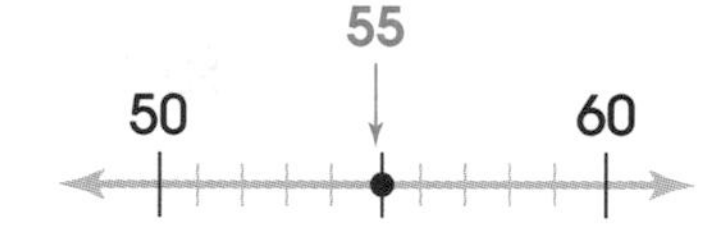

Rounding up makes a larger number. 60 is greater than 55.

- 57 is closer to 60 than to 50.
- 57 rounds up to **60**.

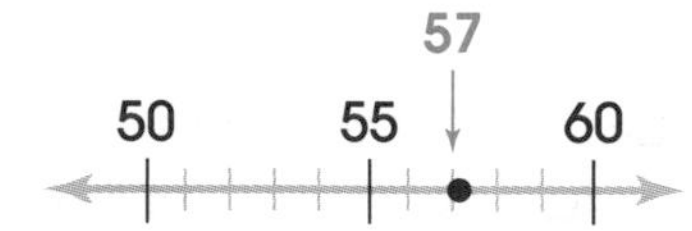

Reminder

Write the rounded numbers on a separate sheet of paper.

Round each number to the nearest ten. You may need to round up or down. Use the number line below for help.

10 20 30 40 50 60 70 80 90

1. 25
2. 52
3. 47
4. 18
5. 85
6. 64

B. You can use a number line to round to the nearest hundred.

- 374 is closer to 400 than to 300.
- 374 rounds up to **400**.

374
300 350 400

- 750 is halfway between 700 and 800.
- 750 rounds up to **800**.

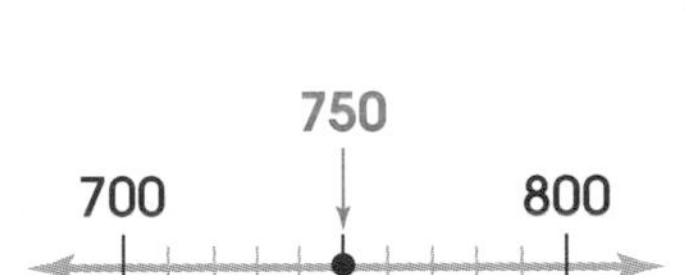

Round each number to the nearest hundred. Use the number line below for help.

100 200 300 400 500 600 700 800 900

1. 220
2. 350
3. 772
4. 435
5. 129
6. 888

C. You can round numbers without using a number line.

Look at the number to the right of the rounding place. If the number is less than 5, round down. If the number is 5 or more, round up. Look at the examples below.

Round **789** to the nearest **ten**.
Look first at the ones place.
The **9** in 789 is more than 5.
789 rounds up to **790**.

Look at the ones place. Round these numbers to the nearest ten.

1. 63
2. 26
3. 347
4. 959
5. 474
6. 9,467

D. You can round any number to any place you need.

To round to the nearest hundred, look at the tens place first.

Round **4,312** to the nearest **hundred**.
Look at the tens place.
The **1** in 4,312 is less than 5.
The number 4,312 rounds down to **4,300**.

Look at the tens place. Round to the nearest hundred.

1. 685
2. 239
3. 357
4. 583
5. 5,891
6. 6,100

Lesson 4 Beginning Addition

Words to Know

addition putting numbers together; to find the *total* amount
sum the amount you get when numbers are added; the *total*
equal (=) the same as

Popcorn for $2 and a movie ticket for $4 costs $6 all together.

You use addition every day. What is one way you might use addition? To figure out the price of a movie and a bag of popcorn, you add the prices together.

A.

Addition means putting numbers together to get a total. The total also may be called the **sum.** Two numbers added together equals the sum. When two amounts are **equal,** they are the same. The symbol = means equal.

two dollars	plus	four dollars	equals	six dollars
$2	+	$4	=	$6

Look at the addition problems below. The same problem is written in two ways. Find the missing words or numbers. Rewrite the addition problems on your paper.

1. five _____ seven _____ twelve
2. 5 + _____ = _____

B.

You know that numbers can be shown as points on a line. A number line can help you add. To add 4 and 5, start by putting your finger on 4. Move your finger 5 spaces to the right. You end up on 9.

4 + 5 = 9

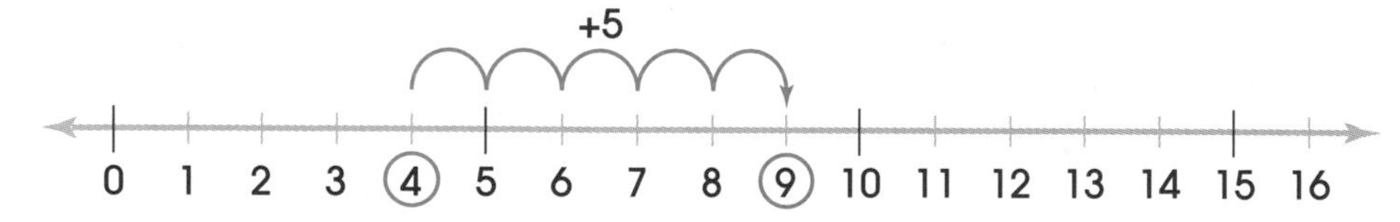

Reminder

Write the sums on a separate sheet of paper.

Use a number line to help you find the sums.

1. $4 + 3$
2. $7 + 1$
3. $5 + 0$

4. $\begin{array}{r} 7 \\ +\,4 \\ \hline \end{array}$
5. $\begin{array}{r} 4 \\ +\,0 \\ \hline \end{array}$
6. $\begin{array}{r} 6 \\ +\,8 \\ \hline \end{array}$
7. $\begin{array}{r} 1 \\ +\,8 \\ \hline \end{array}$
8. $\begin{array}{r} 3 \\ +\,0 \\ \hline \end{array}$

C.

It is not hard to add more than two numbers together. Look at the example below.

Add.

$$\begin{array}{r} 2 \\ 4 \\ 3 \\ +\,7 \\ \hline \end{array}$$

Step 1 Add. 2 + 4 = 6

Step 2 Add. 6 + 3 = 9

Step 3 Add. 9 + 7 = 16

$$\begin{array}{r} 2 \\ 4 \\ 3 \\ +\,7 \\ \hline 16 \end{array}$$

The sum of 2 + 4 + 3 + 7 is 16.

Find the sums below. Show your work.

1. $\begin{array}{r} 6 \\ 2 \\ +\,4 \\ \hline \end{array}$ 2. $\begin{array}{r} 4 \\ 7 \\ +\,5 \\ \hline \end{array}$ 3. $\begin{array}{r} 7 \\ 0 \\ +\,4 \\ \hline \end{array}$ 4. $\begin{array}{r} 1 \\ 3 \\ +\,9 \\ \hline \end{array}$

5. $\begin{array}{r} 4 \\ 8 \\ 3 \\ +\,1 \\ \hline \end{array}$ 6. $\begin{array}{r} 8 \\ 2 \\ 0 \\ +\,6 \\ \hline \end{array}$ 7. $\begin{array}{r} 6 \\ 3 \\ 4 \\ +\,2 \\ \hline \end{array}$ 8. $\begin{array}{r} 4 \\ 6 \\ 1 \\ +\,2 \\ \hline \end{array}$

Reminder

Copy the problems carefully onto a separate sheet of paper.

D.

A place value chart can help you see how larger numbers are added. This can be done in three simple steps. Look at the example below.

Add.

$$\begin{array}{r} 347 \\ +\,251 \\ \hline \end{array}$$

Step 1 Add the digits in the ones place.

Step 2 Add the digits in the tens place.

Step 3 Add the digits in the hundreds place.

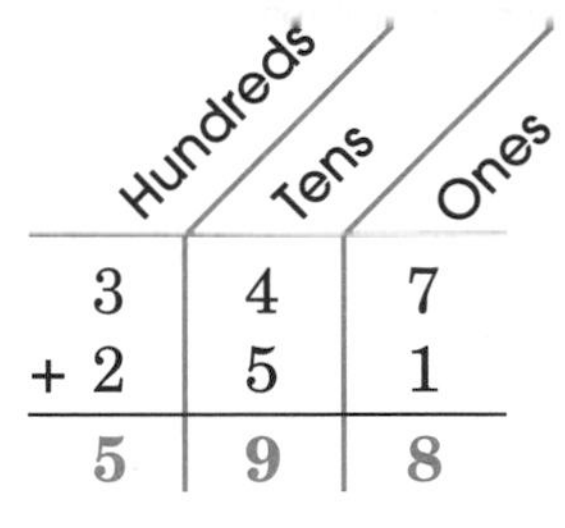

Hundreds	Tens	Ones
3	4	7
+ 2	5	1
5	9	8

The sum of 347 + 251 is 598.

Find the sums below. Show your work.

1. $\begin{array}{r} 60 \\ +\,20 \\ \hline \end{array}$ 2. $\begin{array}{r} 40 \\ +\,45 \\ \hline \end{array}$ 3. $\begin{array}{r} 72 \\ +\,24 \\ \hline \end{array}$ 4. $\begin{array}{r} 372 \\ +\,425 \\ \hline \end{array}$

Lesson 5 Adding with Regrouping

Words to Know

rename to write a number in a different way

regroup to rename a number in order to add, subtract, multiply, or divide

You have already learned how to **rename** a number. You show numbers in different ways by renaming.

A. Renaming numbers is an important skill. Look at the example below. It shows two ways to rename 23.

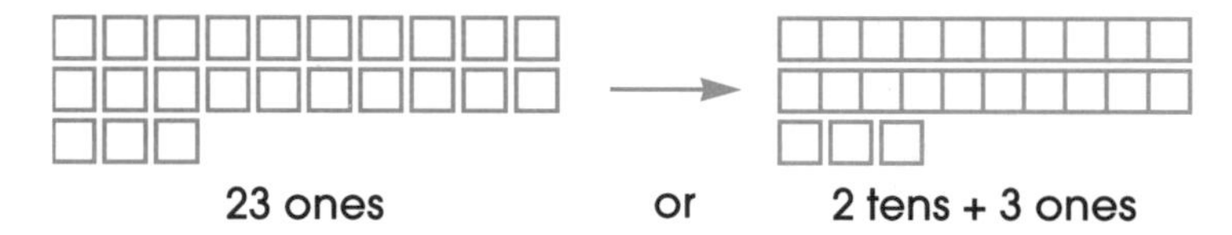

Rename the numbers below.

1. 53 = _____ ones or 53 = _____ tens and _____ ones
2. 79 = _____ ones or 79 = _____ tens and _____ ones
3. 30 = _____ ones or 30 = _____ tens and _____ ones

B. Often you will need to rename when you add numbers. Renaming numbers can make adding easier. When you rename to add, you **regroup** the number so adding is easier. Suppose you want to add 26 and 17. To find the sum, you must regroup. It can be done in four steps. Look at the example below.

Add.
$$\begin{array}{r} 26 \\ +\ 17 \\ \hline \end{array}$$

Step 1 Add the digits in the ones place.
6 ones + 7 ones = 13 ones

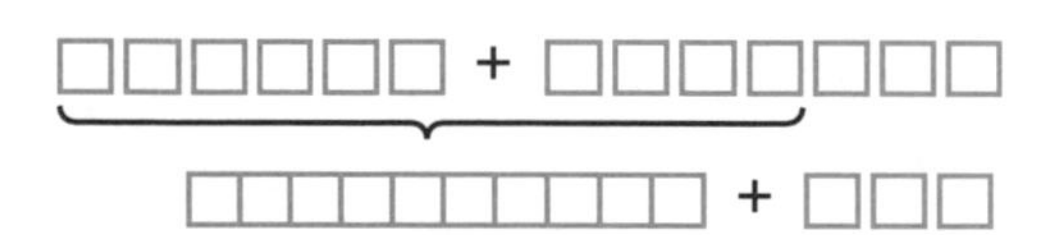

Step 2 Regroup 13 ones.
1 ten + 3 ones

Step 3 Write 3 in the **ones** place.
Write 1 above the **tens** place.

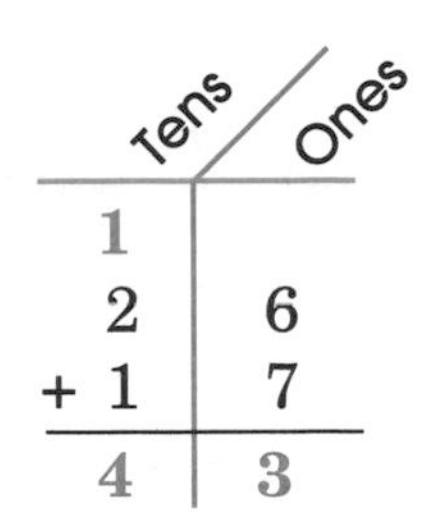

Step 4 Add the digits in the tens place.
$1 + 2 + 1 = 4$
Write 4 in the **tens** place.

The sum of 26 + 17 is 43.

Find the sums below. Regroup when you need to.

1.	2.	3.	4.	5.
64	35	53	27	31
+ 26	+ 49	+ 38	+ 45	+ 29

Reminder

Show your work on a separate sheet of paper.

C.

Sometimes you need to regroup more than once in the same problem. Look at the problem below. Notice how the numbers are regrouped.

Add.

557
+ 389

Step 1 Add the numbers in the ones place. $7 + 9 = 16$

Step 2 Regroup 16 ones.
1 ten + 6 ones
Write 6 in the **ones** place.
Write 1 above the **tens** place.

Hundreds	Tens	Ones
	1	
5	5	7
+ 3	8	9
		6

Step 3 Add the numbers in the tens place.
$1 + 5 + 8 = 14$

Step 4 Regroup 14 (in the tens place).
1 hundred and 4 tens
Write 4 in the **tens** place.
Write 1 above the **hundreds** place.

Hundreds	Tens	Ones
1	1	
5	5	7
+ 3	8	9
	4	6

Step 5 Add the numbers in the hundreds place.
$1 + 5 + 3 = 9$
Write 9 in the **hundreds** place.

Hundreds	Tens	Ones
1	1	
5	5	7
+ 3	8	9
9	4	6

The sum of 557 + 389 is 946.

Find the sums below. Regroup when you need to.

1.	2.	3.	4.	5.
268	787	615	363	540
+ 638	+ 145	+ 325	+ 286	+ 78

Lesson 6 Beginning Subtraction

Words to Know

subtraction taking away one number from another; finding the difference

difference the amount you get when one number is subtracted from another

The word *subtraction* sounds like sub-TRAK-shuhn.

We use subtraction every day. We use it to figure change with money. We may use it to compare prices. What is the price difference between a $5 lunch and a $3 lunch?

A. **Subtraction** means taking away one number from another. When you subtract you find the **difference** between two numbers. Look at the example below. The subtraction problem is 5 – 3 = 2. The difference is $2.

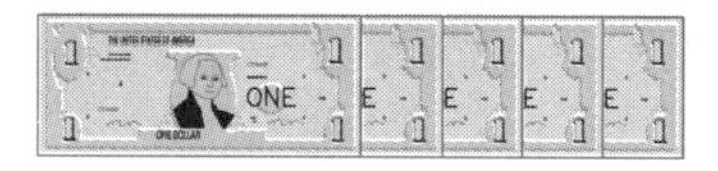

five dollars	minus	three dollars	equals	two dollars
$5	–	$3	=	$2

A number line can help you subtract. Look at the example below.

9 – 4 = 5

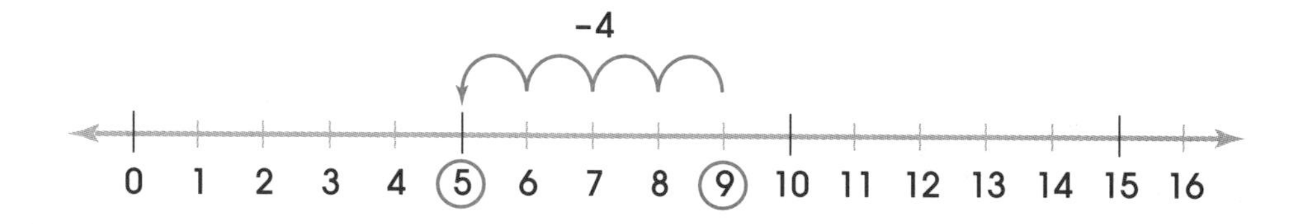

Use a number line to find the differences below.

1. 4 – 1
2. 7 – 2
3. 5 – 3
4. 9 – 5
5. $\begin{array}{r} 8 \\ -\,4 \\ \hline \end{array}$
6. $\begin{array}{r} 5 \\ -\,0 \\ \hline \end{array}$
7. $\begin{array}{r} 9 \\ -\,6 \\ \hline \end{array}$
8. $\begin{array}{r} 7 \\ -\,7 \\ \hline \end{array}$

B.

A place value chart can help you see how large numbers are subtracted. This can be done in 3 simple steps. Look at the example below. Find the difference.

Subtract. 847
− 624

Step 1 Subtract the digits in the ones place.

Step 2 Subtract the digits in the tens place.

Step 3 Subtract the digits in the hundreds place.

Hundreds	Tens	Ones
8	4	7
− 6	2	4
2	2	3

Find the differences below.

1. 50 − 20
2. 65 − 23
3. 89 − 34
4. 74 − 40
5. 588 − 254
6. 774 − 324
7. 551 − 231
8. 654 − 123

Reminder

Show your work on a separate sheet of paper.

C.

You can check your answer to a subtraction problem by using addition. This can be done in two steps.

Subtract. 15
− 8

Step 1 Find the difference.

15
− 8
7

Step 2 Add the difference to the number just above it. The sum should equal the number on the top.

7
+ 8
15

Find the differences below. Then add to check your answers.

1. 69 − 24
2. 25 − 11
3. 39 − 26
4. 685 − 424
5. 669 − 325
6. 839 − 616

Lesson 7 Subtraction with Regrouping

Words to Know

rename to write a number in a different way

regroup to rename a number in order to add, subtract, multiply, or divide

As in addition, you may need to **rename** or **regroup** to subtract easily.

A. You know that the number 23 can be renamed as 2 tens plus 3 ones.

23 = □□□□□□□□□□ □□□□□□□□□□ + □□□

23 = 2 tens + 3 ones

You can also regroup 23 another way. Look at the pictures below. See how 23 can be regrouped to show more ones.

23 = □□□□□□□□□□ + □□□□□□ □□□□□□□

23 = 1 ten + 13 ones

Look at the problem below. See how the top number, 53, has been regrouped to show more ones.

Subtract. 53
−38

Step 1 Can you subtract 8 from 3? No. Regroup 53 to show more ones.

Tens	Ones
4	13
~~5~~	~~3~~
−3	8

Step 2 Subtract the digits in the ones place.

13 – 8 = 5

Write **5** in the **ones** place.

Tens	Ones
4	13
~~5~~	~~3~~
– 3	8
	5

Step 3 Subtract the digits in the tens place.

4 – 3 = 1

Write **1** in the **tens** place.

Tens	Ones
4	13
~~5~~	~~3~~
– 3	8
1	5

The difference between 53 and 38 is 15.

Solve the problems below. When you need to, regroup the numbers in the ones place.

1.	2.	3.	4.
73 – 14	65 – 47	44 – 29	75 – 9

5.	6.	7.	8.
96 – 59	43 – 35	68 – 33	58 – 29

B.

In some problems, you must regroup to show more tens. This can be done in three steps. Look at the example below.

Subtract. 848
– 266

Step 1 Subtract the digits in the ones place.

Step 2 Look at the tens place. Regroup to show more tens.

Step 3 Subtract the digits in the tens and hundreds places.

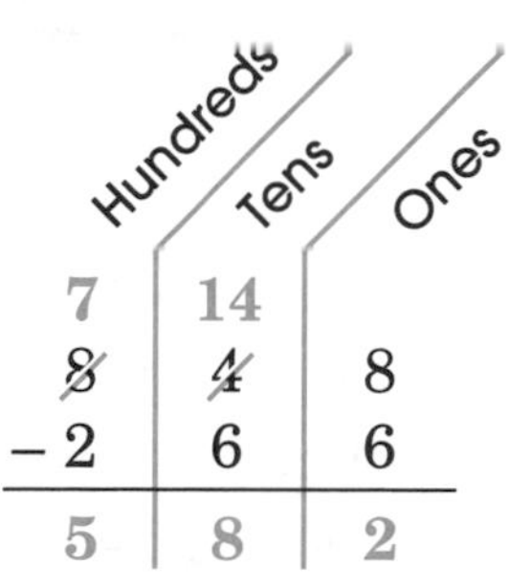

The difference between 848 and 266 is 582.

Solve the subtraction problems below. Regroup to show more tens.

1.	2.	3.	4.
329 – 48	431 – 60	824 – 74	548 – 81

5.	6.	7.	8.
626 – 233	411 – 180	658 – 368	529 – 463

Lesson 8 Subtraction with More Than One Regrouping

Words to Know

rename to write a number in different way

regroup to rename a number in order to add, subtract, or multiply

In larger subtraction problems, you may need to **rename** or **regroup** two or more times.

A.

You can work any subtraction problem one step at a time. Sometimes you will need more ones, tens, or hundreds. Then you will need to regroup. Five steps help you solve larger subtraction problems. Follow the five steps to work the problem below.

Subtract. $\begin{array}{r} 854 \\ -\,567 \\ \hline \end{array}$

Step 1 Can you subtract 7 from 4?
No. Regroup to show more ones.

Step 2 Subtract the digits in the ones place.
14 – 7 = 7

$\begin{array}{r} {\scriptstyle 4\,14} \\ 8\not{5}\not{4} \\ -\,567 \\ \hline 7 \end{array}$

Step 3 Can you subtract 6 from 4?
No. Regroup to show more tens.

$\begin{array}{r} {\scriptstyle 14} \\ {\scriptstyle 7\,\not{4}\,14} \\ \not{8}\not{5}\not{4} \\ -\,567 \\ \hline 7 \end{array}$

Step 4 Subtract the digits in the tens and hundreds place.
tens: 14 – 6 = 8
hundreds: 7 – 5 = 2

$\begin{array}{r} {\scriptstyle 14} \\ {\scriptstyle 7\,\not{4}\,14} \\ \not{8}\not{5}\not{4} \\ -\,567 \\ \hline 287 \end{array}$

Step 5 Add to check your answer.

$\begin{array}{r} 287 \\ +\,567 \\ \hline 854 \end{array}$

The difference between 854 and 567 is 287.

Find the differences below. Regroup when you need to. Check your answers by adding.

1. $\begin{array}{r} 435 \\ -\ 258 \\ \hline \end{array}$
2. $\begin{array}{r} 952 \\ -\ 379 \\ \hline \end{array}$
3. $\begin{array}{r} 356 \\ -\ 178 \\ \hline \end{array}$
4. $\begin{array}{r} 961 \\ -\ 382 \\ \hline \end{array}$
5. $\begin{array}{r} 558 \\ -\ 149 \\ \hline \end{array}$
6. $\begin{array}{r} 422 \\ -\ 48 \\ \hline \end{array}$
7. $\begin{array}{r} 415 \\ -\ 239 \\ \hline \end{array}$
8. $\begin{array}{r} 249 \\ -\ 178 \\ \hline \end{array}$

B.

When you subtract with zeros, you may have to regroup more than once. Follow the example below.

Subtract. $\begin{array}{r} 506 \\ -\ 287 \\ \hline \end{array}$

Look at the problem. There are not enough ones to subtract in the ones column. There are no tens to borrow.

Step 1 Regroup to show more tens.

$\begin{array}{r} {\scriptstyle 4\,10} \\ \not{5}\not{0}6 \\ -\ 287 \\ \hline \end{array}$

Step 2 Now you can regroup to show more ones.

$\begin{array}{r} {\scriptstyle 9} \\ {\scriptstyle 4\,\not{10}\,16} \\ \not{5}\not{0}\not{6} \\ -\ 287 \\ \hline \end{array}$

Step 3 Subtract.

$\begin{array}{r} {\scriptstyle 9} \\ {\scriptstyle 4\,\not{10}\,16} \\ \not{5}\not{0}\not{6} \\ -\ 287 \\ \hline 219 \end{array}$

The difference between 506 and 287 is 219.

Find the differences below. Regroup when you need to. Check your answers by adding.

Reminder

Show your work on a separate sheet of paper.

1. $\begin{array}{r} 203 \\ -\ 56 \\ \hline \end{array}$
2. $\begin{array}{r} 230 \\ -\ 69 \\ \hline \end{array}$
3. $\begin{array}{r} 509 \\ -\ 168 \\ \hline \end{array}$
4. $\begin{array}{r} 305 \\ -\ 138 \\ \hline \end{array}$
5. $\begin{array}{r} 860 \\ -\ 426 \\ \hline \end{array}$
6. $\begin{array}{r} 608 \\ -\ 398 \\ \hline \end{array}$
7. $\begin{array}{r} 620 \\ -\ 185 \\ \hline \end{array}$
8. $\begin{array}{r} 505 \\ -\ 235 \\ \hline \end{array}$
9. $\begin{array}{r} 8{,}020 \\ -\ 3{,}561 \\ \hline \end{array}$

Lesson 9 Multiplication Facts

Words to Know

multiply to add a number one or more times
multiplication multiplying two or more numbers
factor one of the numbers multiplied to find a product
product the answer to a multiplication problem

People multiply every day. You might see that a sandwich costs $2. Suppose you want to buy three. How much will three sandwiches cost? To find out, you can multiply 3 times 2. Three sandwiches will cost $6.

One sandwich costs $2.
Three sandwiches cost $6.

The word *multiplication* sounds like mul-tih-plih-KAY-shuhn.

A.

Multiplying two numbers is faster than adding the same number many times. To **multiply** is to add a number one or more times.

$2 + $2 + $2 = $6

The problem has three groups of three. Three times two equals six. $3 \times 2 = 6$

When you multiply two numbers you are doing **multiplication.** You can write multiplication problems two ways. In both problems below, 3 and 5 are the **factors.** They are the numbers being multiplied. The result, 15, is the **product.**

5	×	3	=	15
five	times	three	equals	fifteen

$$\begin{array}{r} 5 \\ \times\ 3 \\ \hline 15 \end{array}$$

5 ← factor
3 ← factor
15 ← product

Look at the addition problems below. Rewrite each as a multiplication problem. Write each problem in two ways.

1. $1 + 1 + 1 + 1 = 4$
2. $5 + 5 = 10$
3. $4 + 4 + 4 = 12$
4. $2 + 2 + 2 + 2 = 8$
5. $6 + 6 + 6 = 18$

B.

Notice that you can multiply factors in any order. The product will be the same.

Four groups of 2:

$4 \times 2 = 8$ ★★ ★★ ★★ ★★

Two groups of 4:

$2 \times 4 = 8$ ★★★★ ★★★★

Use a multiplication chart to help you remember multiplication facts. To use it, look for the two factors you are multiplying. Point to the box where they both meet. That is your product.

$5 \times 6 = 30$

×	1	2	3	4	5	6	7	8	9	10
1	1	2	3	4	5	6	7	8	9	10
2	2	4	6	8	10	12	14	16	18	20
3	3	6	9	12	15	18	21	24	27	30
4	4	8	12	16	20	24	28	32	36	40
5	5	10	15	20	25	30	35	40	45	50
6	6	12	18	24	30	36	42	48	54	60
7	7	14	21	28	35	42	49	56	63	70
8	8	16	24	32	40	48	56	64	72	80
9	9	18	27	36	45	54	63	72	81	90
10	10	20	30	40	50	60	70	80	90	100

Use the chart above to find the products below.

1. 3×4
2. 4×3
3. 6×8
4. $\begin{array}{r} 9 \\ \times 9 \\ \hline \end{array}$
5. $\begin{array}{r} 7 \\ \times 3 \\ \hline \end{array}$

C.

When you multiply any number by 0, the answer is 0. When you multiply any number by 1, the number stays the same.

$7 \times 0 = 0$ $0 \times 7 = 0$

$7 \times 1 = 7$ $1 \times 7 = 7$

Find the products below.

1. $\begin{array}{r} 9 \\ \times 0 \\ \hline \end{array}$
2. $\begin{array}{r} 6 \\ \times 1 \\ \hline \end{array}$
3. $\begin{array}{r} 0 \\ \times 8 \\ \hline \end{array}$
4. $\begin{array}{r} 1 \\ \times 9 \\ \hline \end{array}$
5. $\begin{array}{r} 1 \\ \times 0 \\ \hline \end{array}$

Lesson 10 Multiplying with Regrouping

Words to Know

regroup to rename a number so you can add, subtract, multiply, or divide

product the answer to a multiplication problem

You have learned to multiply one digit times another. These are the multiplication facts. Sometimes you need to multiply larger numbers. As with addition and subtraction, you may need to **regroup.** When you regroup, you rename a number to make the problem easier to solve.

A. Suppose you want to buy three new CDs. Each one costs $15. How much money do you need all together? To find out, you can solve a multiplication problem.

3 CDs × $15 = $?

Reminder

Use your multiplication chart to help with your facts.

Follow the steps below to regroup and solve the problem.

Multiply.

$$\begin{array}{r} 15 \\ \times\ 3 \\ \hline \end{array}$$

Step 1 Multiply the digit in the ones place by 3. 3 × 5 = 15

Step 2 Regroup 15. 1 ten + 5 ones
Write 5 in the ones place.
Write the 1 above the tens place.

$$\begin{array}{r} {}^{1} \\ 15 \\ \times\ 3 \\ \hline 5 \end{array}$$

Step 3 Multiply the digit in the tens place by 3. 3 × 1 = 3

Step 4 Add. 3 + 1 = 4
Write 4 in the tens place.

$$\begin{array}{r} {}^{1} \\ 15 \\ \times\ 3 \\ \hline 45 \end{array} \leftarrow \text{product}$$

The **product** of 3 times 15 is 45. 3 CDs cost $45.

Find the products. Be sure to follow the four steps. Show your work.

1. $\begin{array}{r} 16 \\ \times\ 2 \\ \hline \end{array}$ 2. $\begin{array}{r} 43 \\ \times\ 5 \\ \hline \end{array}$ 3. $\begin{array}{r} 35 \\ \times\ 6 \\ \hline \end{array}$ 4. $\begin{array}{r} 38 \\ \times\ 4 \\ \hline \end{array}$

5. $\begin{array}{r} 25 \\ \times\ 5 \\ \hline \end{array}$ 6. $\begin{array}{r} 27 \\ \times\ 3 \\ \hline \end{array}$ 7. $\begin{array}{r} 58 \\ \times\ 6 \\ \hline \end{array}$ 8. $\begin{array}{r} 37 \\ \times\ 9 \\ \hline \end{array}$

Reminder

Copy each problem carefully onto a separate sheet of paper.

B.

You've learned how to regroup when you multiply by a two-digit number. What if you multiply by a three-digit number? Sometimes you have to regroup more than once. Look at the problem below. Follow the steps for regrouping.

Multiply. $\begin{array}{r} 264 \\ \times\ \ 6 \\ \hline \end{array}$

Step 1 Multiply the digit in the ones place by 6. $6 \times 4 = 24$

Step 2 Regroup 24. 2 tens + 4 ones
Write **4** in the **ones** place.
Write the **2** above the **tens** place.

$\begin{array}{r} {\scriptstyle 2}\ \ \\ 264 \\ \times\ \ 6 \\ \hline 4 \end{array}$

Step 3 Multiply the digit in the tens place by 6. $6 \times 6 = 36$

Step 4 Add. $36 + 2 = 38$
Regroup 38. 3 hundreds + 8 tens
Write **8** in the **tens** place.
Write **3** above the **hundreds** place.

$\begin{array}{r} {\scriptstyle 3\,2}\ \ \\ 264 \\ \times\ \ 6 \\ \hline 84 \end{array}$

Step 5 Multiply the digit in the hundreds place by 6.
$6 \times 2 = 12$

Step 6 Add. $12 + 3 = 15$
Write **5** in the **hundreds** place.
Write **1** in the **thousands** place.

$\begin{array}{r} {\scriptstyle 3\,2}\ \ \\ 264 \\ \times\ \ 6 \\ \hline 1{,}584 \end{array}$

The product of 6 times 264 is 1,584.

Reminder

A place value chart shows the places in a number. Look at the chart below.

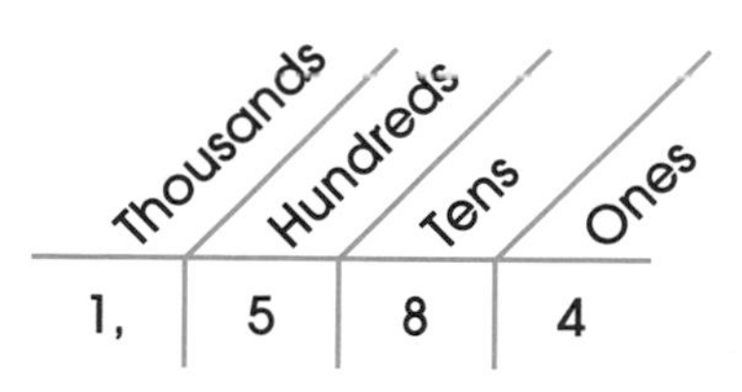

Find the products. Be sure to follow the steps to regroup.

1. $\begin{array}{r} 175 \\ \times\ \ 4 \\ \hline \end{array}$ 2. $\begin{array}{r} 218 \\ \times\ \ 5 \\ \hline \end{array}$ 3. $\begin{array}{r} 348 \\ \times\ \ 3 \\ \hline \end{array}$ 4. $\begin{array}{r} 258 \\ \times\ \ 2 \\ \hline \end{array}$

5. $\begin{array}{r} 369 \\ \times\ \ 7 \\ \hline \end{array}$ 6. $\begin{array}{r} 428 \\ \times\ \ 5 \\ \hline \end{array}$ 7. $\begin{array}{r} 618 \\ \times\ \ 9 \\ \hline \end{array}$ 8. $\begin{array}{r} 854 \\ \times\ \ 8 \\ \hline \end{array}$

Lesson 11 Multiplying Two-Digit Numbers

Words to Know

partial product the number you get when you multiply a number by one digit of another number.

regroup to rename a number so you can add, subtract, multiply, or divide

Multiplication helps us solve everyday problems. Suppose you and 24 friends want to give a party. Each of you brings $12 to pay for the band. How much money would that be all together? To find out fast, you would use multiplication.

A. When you multiply larger numbers, you must take three or more steps. Look at the problem below. Notice how it is solved in three steps.

Multiply. $\begin{array}{r} 24 \\ \times\ 12 \\ \hline \end{array}$

Step 1 Multiply the digit 2 in the ones place by 24.

$2 \times 24 = 48$

This will give a **partial product.** Write 48 under the problem.

$\begin{array}{r} 24 \\ \times\ 12 \\ \hline 48 \end{array}$ ← partial product

The word *partial* sounds like PAHR-shuhl.

Step 2 Multiply the digit 1 in the tens place by 24.

$1 \times 24 = 24$

Write 24 below 48. Put a 0 in the ones place. 240 is the second partial product.

$\begin{array}{r} 24 \\ \times\ 12 \\ \hline 48 \\ 240 \end{array}$ ← partial product, ← partial product

Step 3 Add the partial products.

$48 + 240 = 288$

$\begin{array}{r} 24 \\ \times\ 12 \\ \hline 48 \\ 240 \\ \hline 288 \end{array}$ ← partial product, ← partial product, ← product

The product of 24 times 12 is 288.

Use the three steps to find the products. Show your work.

1. $\begin{array}{r} 42 \\ \times\ 42 \\ \hline \end{array}$ 2. $\begin{array}{r} 81 \\ \times\ 25 \\ \hline \end{array}$ 3. $\begin{array}{r} 11 \\ \times\ 89 \\ \hline \end{array}$ 4. $\begin{array}{r} 72 \\ \times\ 23 \\ \hline \end{array}$

B.

When multiplying two-digit numbers, sometimes you need to **regroup.** Look at the problem below. Notice how it is solved in six steps.

Multiply. $\begin{array}{r} 68 \\ \times\ 25 \\ \hline \end{array}$

Step 1 Start with the 5.
Multiply 5 by 8. $5 \times 8 = 40$

Step 2 Regroup 40. 4 tens + 0 ones.
Write **0** in the ones place.
Write the **4** above the tens place.

$\begin{array}{r} {}^{4} \\ 68 \\ \times\ 25 \\ \hline 0 \end{array}$

Step 3 Multiply 5 by 6 in the tens place.
$5 \times 6 = 30$
Add. $30 + 4 = 34$
Write **3** in the hundreds and **4** in the tens place.

$\begin{array}{r} {}^{4} \\ 68 \\ \times\ 25 \\ \hline 340 \end{array}$

Step 4 Now look at the 2.
Multiply. $2 \times 8 = 16$
Regroup 16. Put a **0** in the ones place.
Write **6** in the tens place.
Write **1** above the tens place.

$\begin{array}{r} {}^{1} \\ 68 \\ \times\ 25 \\ \hline 340 \\ 60 \end{array}$

Step 5 Multiply 2 by 6 in the tens place.
$2 \times 6 = 12$
Add. $12 + 1 = 13$
Write **13** in the hundreds place.

$\begin{array}{r} {}^{1} \\ 68 \\ \times\ 25 \\ \hline 340 \\ 1{,}360 \end{array}$

Step 6 Add the partial products.
$340 + 1{,}360 = 1{,}700$

$\begin{array}{r} 68 \\ \times\ 25 \\ \hline 340 \\ 1{,}360 \\ \hline 1{,}700 \end{array}$

The product of 68 times 25 is 1,700.

Reminder

When you regroup, you rename the number to make the problem easier.

Reminder

Use regrouping when you add the partial products.

$\begin{array}{r} {}^{1} \\ 340 \\ +\ 1360 \\ \hline 1700 \end{array}$

Use the steps to find the products. Show your work.

1. $\begin{array}{r} 42 \\ \times\ 16 \\ \hline \end{array}$ 2. $\begin{array}{r} 56 \\ \times\ 22 \\ \hline \end{array}$ 3. $\begin{array}{r} 38 \\ \times\ 54 \\ \hline \end{array}$ 4. $\begin{array}{r} 87 \\ \times\ 63 \\ \hline \end{array}$

Lesson 12 Multiplying with Zeros

Word to Know

product the answer in a multiplication problem

There is a quick and easy way to multiply numbers by 10, 100, or 1,000.

A. To multiply a number by 10, first write the number. Then write *one* 0 right after the number. This will give you the final result or **product.** Look at the multiplication facts for 10.

$10 \times 1 = 10$	$10 \times 2 = 20$	$10 \times 3 = 30$
$10 \times 4 = 40$	$10 \times 5 = 50$	$10 \times 6 = 60$
$10 \times 7 = 70$	$10 \times 8 = 80$	$10 \times 9 = 90$

These are easy to remember. It's easy to multiply any number times 10, 100, 1,000. Look at these examples.

$73 \times 10 = 730$

To multiply a number by 100, first write the number. Then write *two* 0s after the number.

$73 \times 100 = 7{,}300$

To multiply a number by 1,000, first write the number. Then write *three* 0s.

$73 \times 1{,}000 = 73{,}000$

Find the products below.

1. $\begin{array}{r} 42 \\ \times\ 10 \\ \hline \end{array}$
2. $\begin{array}{r} 83 \\ \times\ 10 \\ \hline \end{array}$
3. $\begin{array}{r} 485 \\ \times\ 100 \\ \hline \end{array}$
4. $\begin{array}{r} 6{,}385 \\ \times\ \ \ 100 \\ \hline \end{array}$
5. 357×100
6. $84 \times 1{,}000$
7. $842 \times 1{,}000$

B. You can multiply other numbers with zeros quickly, too. Look at the problems that follow. You can multiply the long way. Or, you can use a quicker way to get your answer. Look at the example for 6×20.

Longer Way	Quicker Way
$\begin{array}{r} 20 \\ \times \quad 6 \\ \hline 120 \end{array}$	Think $6 \times 2 = 12$. Write **one** zero next to the 12. $6 \times 20 = 120$

What if you have more zeros? The quicker way will work for any number of zeros. $6 \times 2{,}000 = ?$

Reminder

A comma helps you see where the thousands place starts.

Longer Way	Quicker Way
$\begin{array}{r} 2{,}000 \\ \times \quad 6 \\ \hline 12{,}000 \end{array}$	Think $6 \times 2 = 12$ Write **three** zeros next to the 12. $6 \times 2{,}000 = 12{,}000$

Use the quicker way to find the products. Show your work on separate paper. Use the longer way to check.

1. 2×30
2. 4×50
3. 6×70
4. 8×40
5. 3×200
6. 5×400
7. $2 \times 8{,}000$
8. $7 \times 7{,}000$

C.

Suppose you have two-digit numbers to multiply. If there is a 0 in one of the numbers, there is a quicker way to multiply. Look at 30 in the example below. The 0 in the ones place tells you there are no ones. You could multiply 68 by 0 the long way, but you don't need to. A quicker way is to just put a 0 in the ones place. Then multiply the rest of the numbers in the problem.

Longer Way	Quicker Way
$\begin{array}{r} 68 \\ \times \quad 30 \\ \hline 00 \\ 2{,}040 \\ \hline 2{,}040 \end{array}$	$\begin{array}{r} 68 \\ \times \quad 30 \\ \hline 2{,}040 \end{array}$

Reminder

Any number multiplied by 0 equals 0.

Solve the problems. Use the quicker way. Show your work.

1. $\begin{array}{r} 16 \\ \times\ 10 \\ \hline \end{array}$
2. $\begin{array}{r} 91 \\ \times\ 40 \\ \hline \end{array}$
3. $\begin{array}{r} 27 \\ \times\ 90 \\ \hline \end{array}$

Lesson 13 Division Facts

Words to Know

divide find out how many times a number contains another
division dividing
quotient the answer in a division problem

Suppose you and a friend are at a restaurant. The restaurant bill is $10. How much does each of you owe? To find out, you would divide the bill, $10, by two. You each owe $5.

The word *division* sounds like duh-VIHZ-uhn.

A. You **divide** to find out how many times one number contains another. When you divide are using **division.** You and 3 friends earn $20 raking leaves. How much money will each person get? To find out, divide 20 by 4. You each get $5.

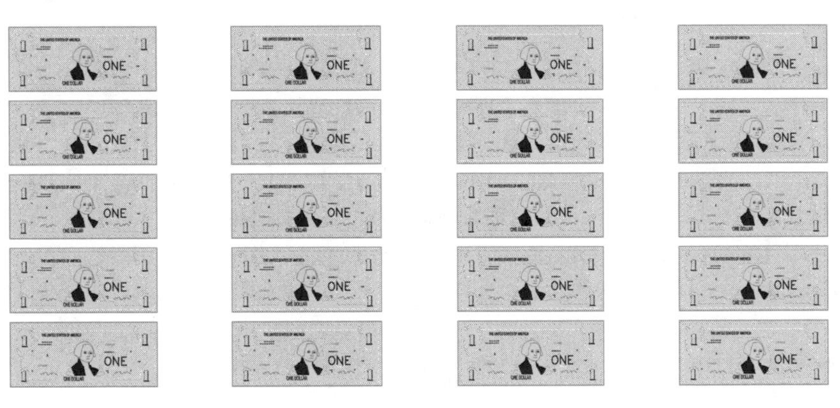

A division problem can be written using the division symbol ÷. This symbol means *divided by.*

20 divided by 4 ⟶ $20 \div 4 = 5$

Twenty divided by four equals five. The answer to the division problem is called the **quotient.**

The word *quotient* sounds like KWO-shuhnt.

Write these division problems in numbers. Use the division symbol.

1. Ten divided by two equals five.
2. Fifteen divided by five equals three.
3. Twelve divided by two equals six.
4. Twenty divided by four equals five.
5. Eighteen divided by three equals six.

B.

Using a chart of multiplication facts can help you divide. Suppose you want to divide 5 into 35. This problem can be solved in three steps using the chart. Follow the steps.

Divide. $35 \div 5$

Step 1 Find the number 5 in the first column.

Step 2 Move your finger across that row until you find 35.

Step 3 Move up that column until you reach the top row. That number is the quotient, or answer. $35 \div 5 = 7$

×	1	2	3	4	5	6	(7)	8	9	10
1	1	2	3	4	5	6	7	8	9	10
2	2	4	6	8	10	12	14	16	18	20
3	3	6	9	12	15	18	21	24	27	30
4	4	8	12	16	20	24	28	32	36	40
(5)	5	10	15	20	25	30	(35)	40	45	50
6	6	12	18	24	30	36	42	48	54	60
7	7	14	21	28	35	42	49	56	63	70
8	8	16	24	32	40	48	56	64	72	80
9	9	18	27	36	45	54	63	72	81	90
10	10	20	30	40	50	60	70	80	90	100

Use the chart above to find the quotients below.

1. $36 \div 6$
2. $18 \div 6$
3. $42 \div 7$
4. $49 \div 7$
5. $72 \div 8$
6. $54 \div 6$
7. $20 \div 2$
8. $56 \div 8$

C.

When you divide any number by itself, the answer is always 1.

$8 \div 8 = 1$　　　$7{,}777 \div 7{,}777 = 1$

When you divide any number by one, the number stays the same.

$5 \div 1 = 5$　　　$9 \div 1 = 9$

Find the quotients for the problems below.

1. $9 \div 9$
2. $4 \div 1$
3. $5 \div 1$
4. $3 \div 3$
5. $25 \div 25$

Lesson 14 Long Division

Words to Know

dividend the number to be divided
divisor the number to divide by
quotient the answer in a division problem

The word *dividend* sounds like DIHV-uh-dehnd.
The word *divisor* sounds like duh-VY-zuhr.

A division problem has three parts. The number to be divided is called the **dividend.** The number to divide by is called the **divisor.** The answer is the **quotient.** You can write a division problem in two ways. Look at the examples below. Writing a divsion problem the second way is helpful when you have to divide larger numbers.

dividend ÷ divisor = quotient

$\text{divisor}\overline{)\text{dividend}}$ with quotient written above

$12 \div 2 = 6$

$2\overline{)12}$ with 6 above

A. Sometimes you have to divide into a divisor that is larger than the numbers in your division chart. You will need to use long division. You can still use the chart to help you divide. Some quotients, or answers, will have more than one digit. Follow the steps for long division.

Divide. $72 \div 3 \longrightarrow 3\overline{)72}$

Step 1 Decide how many 3s there are in 7. There are 2.
Write 2 above the 7 in the dividend.

$3\overline{)72}$ with 2 above

Step 2 Multiply. $2 \times 3 = 6$
Write the 6 below the 7 in the dividend.

$3\overline{)72}$ with 2 above; 6 below

Step 3 Subtract. $7 - 6 = 1$
Bring down the 2. Decide how many 3s there are in 12. There are 4.
Write 4 above the 2 in the dividend.

$3\overline{)72}$ with 24 above; -6 below; 12

Step 4 Multiply. 4 × 3 = 12
Write the **12** below the 12.
Subtract. 12 – 12 = **0**

```
  24
3)72
 - 6
  12
- 12
   0
```

The quotient of 72 divided by 3 is 24.

Follow the long division steps to find the quotients.

1. 2)56 2. 3)87 3. 2)92 4. 6)84

B.

You can use the same long division steps to divide a larger dividend. Look at the example and steps below.

Divide. 365 ÷ 5 ⟶ 5)365

Step 1 Since 3 is smaller than 5, the number 5 does not divide into 3. Decide how many 5s there are in 36. There are 7.
Write **7** above the 36.

```
   7
5)365
```

Step 2 7 × 5 = 35. Write **35** under 36.

```
   7
5)365
  35
```

Step 3 Subtract. 36 – 35 = 1
Bring down the **5**. Decide how many 5s there are in 15. There are 3.
Write the **3** above the 5.

```
   73
5)365
 - 35
   15
```

Step 4 Multiply. 3 × 5 = 15
Write **15** under 15. Subtract.
15 – 15 = **0**

```
   73
5)365
 - 35
   15
 - 15
    0
```

The quotient of 365 divided by 5 is 73.

Checking the answer to a division problem is not hard. Just multiply the quotient by the divisor. The result should be the same as the dividend.

```
            132 ⟵ quotient          132 ⟵ quotient
divisor ⟶ 3)396 ⟵ dividend        ×   3 ⟵ divisor
                                    396 ⟵ dividend
```

Find the quotients. Multiply to check your answers.

1. 5)380 2. 4)632 3. 3)762 4. 2)494 5. 4)448

Lesson 15 Division with Remainders

Words to Know

remainder the number left over in a division problem
quotient the answer in a division problem

Suppose that three friends are at an amusement park. They have 17 ride tickets to share. But 3 does not divide evenly into 17. There are 2 tickets left over. What is left over is called the **remainder.** The remainder is 2.

$$3\overline{)17}\quad \text{5 R2}$$

Reminder

The dividend is the number to be divided. The divisor is the number to divide by.

A. One number does not always divide evenly into another number. Sometimes there is a number left over. This number is called a remainder. The remainder is part of the **quotient.** The remainder can never be larger than the divisor. To solve a problem with a remainder, look at the steps below.

Divide. $3\overline{)95}$

Step 1 Decide how many 3s there are in 9. There are 3.
Write 3 above the 9.

$$\begin{array}{r} 3 \\ 3\overline{)95} \end{array}$$

Step 2 Multiply. 3 × 3 = 9
Write 9 below the dividend

$$\begin{array}{r} 3 \\ 3\overline{)95} \\ 9 \end{array}$$

Step 3 Subtract. 9 – 9 = 0
Bring down the 5.

$$\begin{array}{r} 3 \\ 3\overline{)95} \\ -9 \\ \hline 05 \end{array}$$

Step 4 Decide how many 3s there are in 5.
You know there is at least one 3 because 1 × 3 = 3. But there are not two 3s because 2 × 3 = 6. Put 1 above the 5.

$$\begin{array}{r} 31 \\ 3\overline{)95} \\ -9 \\ \hline 05 \end{array}$$

Step 5 Multiply. $1 \times 3 = 3$

Write **3** under the 5.

```
  31
3)95
 -9
  05
   3
```

Make sure the number you get when you subtract is smaller than your divisor.

Step 6 Subtract. $5 - 3 = 2$

There are no more digits in the dividend to bring down. 2 is your remainder. Write **R2** to show 2 is left over.

```
  31 R2
3)95
 -9
  05
 - 3
   2
```

The quotient for $95 \div 3$ is 31 with a remainder of 2.

Find the quotient. Use R to show a remainder, if there is one. The first problem has been done for you in the margin.

1. $4\overline{)47}$ 2. $2\overline{)443}$ 3. $3\overline{)397}$

4. $6\overline{)664}$ 5. $4\overline{)364}$ 6. $5\overline{)551}$

7. $4\overline{)583}$ 8. $3\overline{)492}$ 9. $6\overline{)735}$

Soluton to problem 1.

```
  11 R3
4)47
 -4
  07
 - 4
   3  <- remainder
```

B.

How can you check an answer with a remainder? Multiply the quotient by the divisor. Then, add your remainder. Your result should be the dividend.

Check. $147 \div 5 = 29$ R2

Step 1 Multiply. 29×5

```
  4
  29
×  5
 145
```

Step 2 Add. $145 + 2 = 147$

Compare 147 to the dividend. $147 = 147$

$147 \div 5 = 29$ R2 is correct.

Check the answers to the problems below. If an answer is wrong, find the correct answer.

1. $51 \div 4 = 12$ R3 2. $78 \div 3 = 15$ 3. $105 \div 4 = 26$ R1

4. $235 \div 3 = 78$ R4 5. $277 \div 5 = 55$ R2 6. $789 \div 6 = 130$ R5

Find the quotients. Check your answers

1. $7\overline{)289}$ 2. $5\overline{)230}$ 3. $9\overline{)468}$ 4. $7\overline{)342}$

Lesson 16 Division with Larger Numbers

Words to Know

quotient the answer in a division problem
remainder the number left over in a division problem

Sometimes you will need to divide very large numbers. What if you want to split up $350 among 25 people? How much money will each person get? The **quotient** answers the question. Each person will get $14.

A.

You can divide by numbers with more than one digit. Follow the example below.

Divide. $32\overline{)235}$

Step 1 Decide where to put the first digit of the quotient. 32 is bigger than 2. 32 is bigger than 23. 32 is less than 235. Put the first digit above the 5 in the dividend.

$$32\overline{)235}$$

Step 2 Divide 235 by 32. Hint: Round 32 to 30. Look at the 3 in 30. Decide how many 3s there are in 23. There are 7 with a **remainder.** Write 7 above the 5 in the dividend.

$$\begin{array}{r} 7 \\ 32\overline{)235} \end{array}$$

Step 3 Multiply. $7 \times 32 = 224$
Write 224 below 235.

$$\begin{array}{r} 7 \\ 32\overline{)235} \\ 224 \end{array}$$

Step 4 Subtract. $235 - 224 = 11$
There are no more digits to bring down. 11 is smaller than 32. 11 is the remainder. Write R11 next to the 7.

$$\begin{array}{r} 7\text{ R11} \\ 32\overline{)235} \\ -\,224 \\ \hline 11 \end{array}$$

The quotient of 235 divided by 32 is 7 R11.

Find the quotients. Show all of your work.

1. $26\overline{)298}$	2. $75\overline{)609}$	3. $56\overline{)463}$
4. $52\overline{)312}$	5. $49\overline{)160}$	6. $19\overline{)148}$
7. $16\overline{)432}$	8. $38\overline{)904}$	9. $67\overline{)144}$
10. $11\overline{)697}$	11. $49\overline{)735}$	12. $53\overline{)858}$

B. Division takes many steps. When you round to decide the quotient, you can make a mistake. The quotient may be too big or too small. You can check for mistakes as you go. Look at the examples below.

Quotient Too Small	Try A Bigger Quotient
$37\overline{)229}$ quotient 5; -185; 44 ← remainder is bigger than divisor	$37\overline{)229}$ quotient 6; -222; 7 ← remainder is less than divisor

Quotient Too Big	Try A Smaller Quotient
$54\overline{)421}$ quotient 8; -432 ← product is bigger than dividend	$54\overline{)421}$ quotient 7; -378 ← product is less than dividend

Find the quotients. Stop and check for mistakes as you go.

1. $45\overline{)855}$	2. $24\overline{)231}$	3. $51\overline{)920}$
4. $79\overline{)948}$	5. $98\overline{)435}$	6. $62\overline{)703}$
7. $23\overline{)437}$	8. $46\overline{)546}$	9. $58\overline{)680}$
10. $15\overline{)789}$	11. $36\overline{)994}$	12. $17\overline{)782}$

Reminder

Be careful when you multiply. Make sure the number you get after you subtract is less than your divisor.

Lesson 17 Common Factors

Words to Know

factor one of the numbers multiplied to find a product

common factors factors that two products have that are the same

greatest common factor (GCF) the largest factor that two or more products have that are the same

Number factors are not new to you. You already use factors when you multiply and divide.

A.

The numbers you multiply to get a product are called **factors.** Factors of a number divide into the number evenly. No remainder is left. Look at the factors of 12. Notice that six numbers divide into 12 evenly.

$12 \div 1 = 12$ $\quad 12 \div 2 = 6$ $\quad 12 \div 3 = 4$

$12 \div 4 = 3$ $\quad 12 \div 5 = 2$ R2 $\quad 12 \div 6 = 2$ $\quad 12 \div 12 = 1$

The factors of 12 are 1, 2, 3, 4, 6, 12. 5 is not a factor because $12 \div 5$ has a remainder.

Reminder

1 and the number itself are always factors.

You can find all the factors of a number by dividing. 1 is always a factor of a number. The number itself is always a factor, too. Follow the example below to find the factors of 15. You know 1 and 15 are factors. Divide to find other factors. Start with 2.

$$\begin{array}{r} 7 \text{ R1} \\ 2\overline{)15} \\ -14 \\ \hline 1 \end{array} \qquad \begin{array}{r} 5 \\ 3\overline{)15} \\ -15 \\ \hline 0 \end{array} \qquad \begin{array}{r} 3 \text{ R3} \\ 4\overline{)15} \\ -12 \\ \hline 3 \end{array}$$

Stop. 5 is next. You already know 5 is a factor from dividing by 3. The factors of 15 are 1, 3, 5, and 15.

Find all the factors of the following numbers.

1. Find the three factors of 4.
2. Find the four factors of 8.
3. Find the three factors of 9.
4. Find the six factors of 20.

B.

Many numbers have **common factors.** This means they have a factor that is the same. The factors of 6 and 12 are listed below. Look closely. What factors do 6 and 12 have in common?

6: **1**, **2**, **3**, **6**

12: **1**, **2**, **3**, 4, **6**, 12

6 and 12 have four common factors: 1, 2, 3, 6.

8: **1**, 2, 4, 8

15: **1**, 3, 5, 15

8 and 15 have one common factor: 1.

List the factors for the numbers below. Name the common factors in each pair.

1. 4 and 8
2. 9 and 12
3. 5 and 15
4. 4 and 10

Reminder

Factors are the numbers you multiply to find a product.

C.

The **greatest common factor (GCF)** is the largest factor that two or more products share. To find the greatest common factor of two or more numbers, use these three steps. Follow the example below.

What is the greatest common factor of 12 and 18?

Step 1	Write the numbers.	**12** **18**
Step 2	Find the factors of each number.	12: 1, 2, 3, 4, 6, 12 18: 1, 2, 3, 6, 9, 18
Step 3	Find the largest factor that is the same for both. This is the greatest common factor. The GCF of 12 and 18 is **6**.	12: 1, 2, 3, 4, **6**, 12 18: 1, 2, 3, **6**, 9, 18

List the factors of each number below. Name the greatest common factor in each pair.

1. 6 and 15
2. 6 and 9
3. 3 and 10
4. 12 and 24
5. 8 and 16
6. 2 and 5
7. 15 and 20
8. 18 and 30

Lesson 18 Least Common Multiple

Words to Know

multiple the product of multiplying a number by other whole numbers

compare to look at two things and see how they are the same or different

least common multiple (LCM) the smallest multiple that two or more numbers share that is not 0

The word *multiple* sounds like MUL-tuh-puhl.

You already know how to find multiples of a number. When you multiply two whole numbers, the product is a **multiple** of each number.

A. In this lesson you will learn to find different multiples of numbers. Zero is always the first multiple of any number. The example below shows multiples of 3.

Reminder

The result of multiplication is called a product.

$0 \times 3 = 0$ $1 \times 3 = 3$

$2 \times 3 = 6$ $3 \times 3 = 9$ $4 \times 3 = 12$ ...

The first 5 multiples of 3 are 0, 3, 6, 9, and 12.

List the first seven multiples of each number. Don't forget to begin with zero.

1. 3	2. 4	3. 5	4. 6
5. 7	6. 8	7. 9	8. 10

B. Now **compare** the multiples of 4 and 6 to see how they are the same or different. Only the first seven multiples are listed below. But a list of multiples could go on and on.

4: 0, 4, 8, 12, 16, 20, 24

6: 0, 6, 12, 18, 24, 30, 36

Notice that some of the common multiples of 4 and 6 are 0, 12, and 24.

The smallest multiple that two numbers have in common that is not zero is called the **least common multiple,** or **LCM.**

Look again at the multiples of 4 and 6 in the example. Then answer each question.

1. In the list below, which number is a common multiple of 4 and 6?
   a. 8
   b. 18
   c. 24
2. In the list below, which number is not a common multiple of 4 and 6.
   a. 0
   b. 18
   c. 12
3. Not counting zero, what is the least common multiple of 4 and 6?
   a. 6
   b. 12
   c. 4

Reminder

Do not count 0 as a least common multiple (LCM).

C.

What is the smallest multiple that 2 and 5 have in common? To find the least common multiple of any two numbers, follow the three steps below.

Step 1 List the first few multiples of each number.

2: 0, 2, 4, 6, 8, 10
5: 0, 5, 10, 15

Step 2 Compare the two lists of multiples. Find the smallest multiple, except 0, that is in both lists. The LCM of 2 and 5 is **10**.

2: 0, 2, 4, 6, 8, **10**
5: 0, 5, **10**, 15

Step 3 Sometimes you cannot find a common multiple right away. You may have to list more multiples in order to find one.

List multiples for each pair of numbers below. Name the least common multiple (LCM) of each pair.

1. 3 and 4	2. 4 and 5	3. 6 and 8	4. 6 and 9
5. 2 and 3	6. 4 and 9	7. 3 and 8	8. 4 and 8
9. 3 and 5	10. 1 and 5	11. 2 and 7	12. 6 and 4

Lesson 19 Whole Numbers: Problem Solving

Words to Know

addition clue words words or groups of words that tell you to add

subtraction clue words words or groups of words that tell you to subtract

multiplication clue words words or groups of words that tell you to multiply

division clue words words or groups of words that tell you to divide

arithmetic operations addition, subtraction, multiplication, and division

Clue words are important. They can help you decide how to solve a math problem. Read the clue words carefully.

addition clue words	in all, both, total, all together, together
subtraction clue words	how many more, greater, difference, how much more, how many less, fewer, remain
multiplication clue words	at $4 each, for 3 days, in 6 weeks, per hour
division clue words	into how many, how many times, how much/many did each

A. Addition, subtraction, multiplication, and division are called **arithmetic operations.** Clue words help you know which operation to use. Look at the example below.

Reminder

Not all problems have clue words.

Jake has 87 baseball cards. Carl has 59.
How many more cards does Jake have?

Clue words: how many more
Operation: subtraction

Read each problem below. Find the clue words. Name the operation to use.

1. Shane baby-sat for 6 hours. At $2 per hour, how much money did Shane earn?
2. Kim won 9 swimming medals this season. Christy won 12. How many medals did both girls win all together?
3. Pat and Fred practiced piano the same number of hours. Together they practiced 98 hours. How many hours did each boy practice?

B. Five simple steps can help you solve math problems. Read the problem and the steps below. Notice how the steps work.

Curtis served 107 hamburgers on Saturday. On Sunday, he served 98 hamburgers. How many hamburgers did Curtis serve in all?

Step 1	Read the problem.	
Step 2	Learn what you must find out.	How many hamburgers did Curtis serve in all?
Step 3	Notice the clue words. Decide which operation to use.	In all tells you to add.
Step 4	Write and solve the problem.	$107 + 98$ = **205 hamburgers**
Step 5	Check your work. Does the answer make sense?	Yes; 205 is more than 107 and 98.

Follow the five steps to answer the questions below. Then solve the problems.

1. Hector has 75 candy bars to sell. Today, he sold 36. How many candy bars does he have left to sell?
   a. What are the clue words?
   b. Which operation will you use?
   c. Solve the problem.
2. The drama club sold $435 worth of tickets to the spring play. 87 tickets were sold. How much did each ticket cost?
   a. What are the clue words?
   b. Which operation will you use?
   c. Solve the problem.
3. Marty earns $48 a month working Saturdays. How much does he earn in a year? (There are 12 months in a year.)
   a. What are the clue words?
   b. Which operation will you use?
   c. Solve the problem.

Unit 2 Fractions

Lesson 20 What Is a Fraction?

Words to Know

fraction a number that names part of a whole

numerator the top number in a fraction; the numerator tells how many parts are being used

denominator the bottom number in a fraction; the denominator tells how many parts are in the whole

The word *fraction* sounds like FRAK-shuhn.

A **fraction** is a part of a whole. When people measure something, they often use fractions. They talk about a $\frac{1}{2}$ pound, $\frac{1}{4}$ cup, or $\frac{5}{8}$ inch. These are all fractions.

A. Look at the square below. The whole square is divided into 4 parts. Each part is the same size. Notice that 3 of the 4 parts are shaded. This picture shows three-fourths.

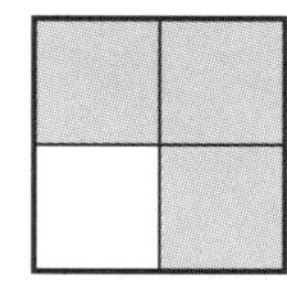

You can write a fraction to show how many parts are shaded. Look at the example below.

Step 1	Write the total number of parts as the bottom number.	$\frac{\ }{4}$
Step 2	Write the number that is shaded as the top number.	$\frac{3}{4}$

The word *denominator* sounds like dee-NAHM-ih-nayt-er. The word *numerator* sounds like NOO-mer-ayt-er.

Every fraction has a **numerator** and a **denominator.** The denominator is the bottom number. It tells how many parts are in the whole. The numerator is the top number. It tells how many parts are being used. See the examples that follow.

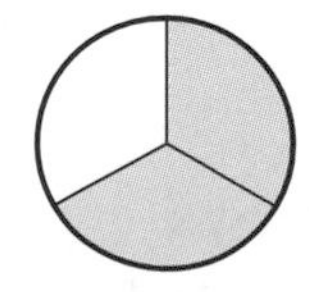

$\frac{2}{3}$ ← numerator ← denominator

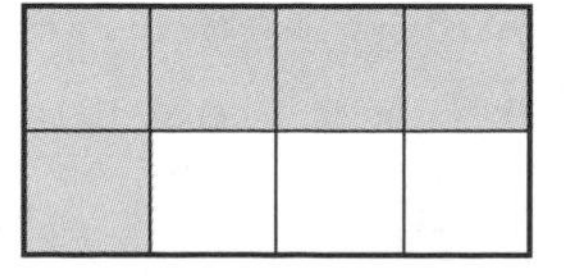

$\frac{5}{8}$ ← numerator ← denominator

The fraction $\frac{2}{3}$ is said as two-thirds. The fraction $\frac{5}{8}$ is said as five-eighths.

Look at the shapes below. Count the total number of parts in each shape. Write a fraction to tell what part of the whole shape is shaded.

1.　　2.　　3.

4.　　5.　　6.

Reminder

Draw the shapes on a separate sheet of paper. Write the fraction.

B. You also can use fractions to name part of a group. Look at the group of circles below. There are five circles. Four of the five circles are shaded. It shows four-fifths. What fractions of the squares are shaded?

$\frac{4}{5}$ four-fifths　　$\frac{3}{6}$ three-sixths

Look at the groups below. Write a fraction for the shaded parts.

1.　　2.

3.　　4.

C. Sometimes a fraction names a whole. Look at the rectangle below. There are six equal parts. All six parts are shaded.

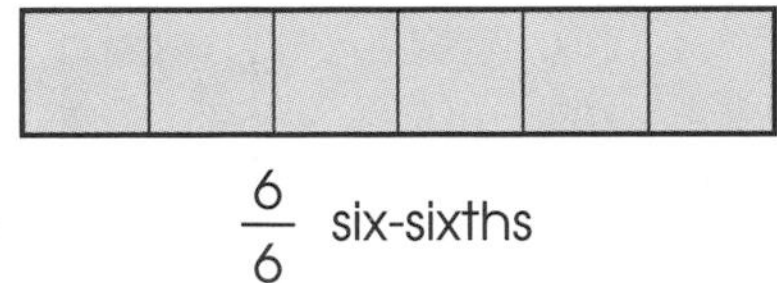

$\frac{6}{6}$ six-sixths

Look at the shapes below. Write a fraction for each whole.

1.　　2.　　3.　　4.

Lesson 21 Equivalent Fractions

Words to Know

equivalent fractions fractions that name the same amount

When you cook, you often use fractions. Sometimes you will need to find equal fractions. You can use $\frac{2}{4}$ cup of sugar or $\frac{1}{2}$ cup of sugar. The two amounts are the same.

A.

Equivalent fractions are equal. They have different numbers, but they name the same amount. Each circle below is the same size, but each is divided into a different number of parts. Look at the three circles. Think about these questions:

1. How many parts are in each circle?
2. How many parts of each circle are shaded?
3. In each circle, is an *equivalent* amount shaded?

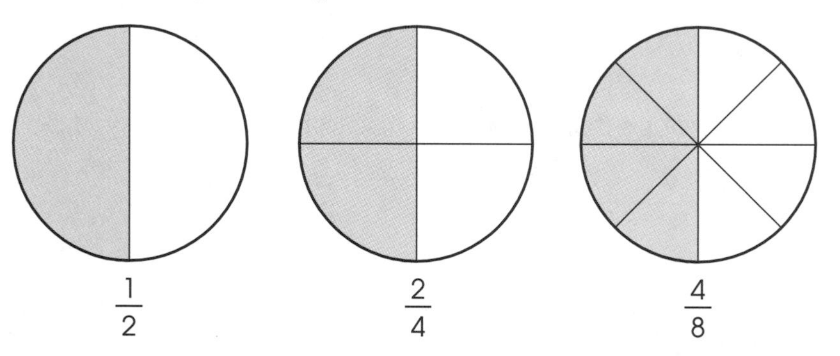

Half of each circle is shaded. All three fractions have the same value. They are equivalent.

$\frac{1}{2} = \frac{2}{4} = \frac{4}{8}$

Reminder

Copy the fraction pairs onto your paper. Write *yes* or *no* next to each pair.

Look at each pair of shapes. Write *yes* if they show equivalent fractions. Write *no* if they are not equivalent.

1. $\frac{1}{4}$ $\frac{1}{3}$

2. $\frac{4}{4}$ 1

3. 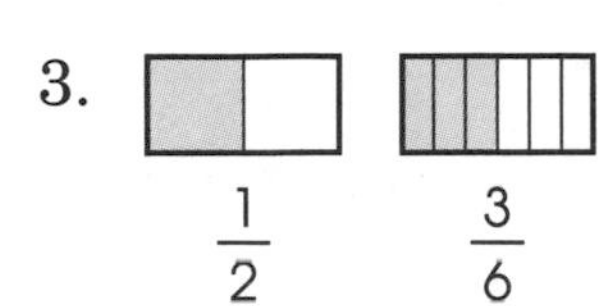

4. 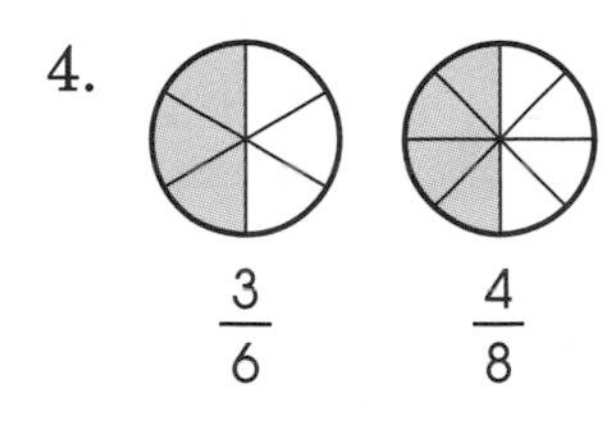

5. 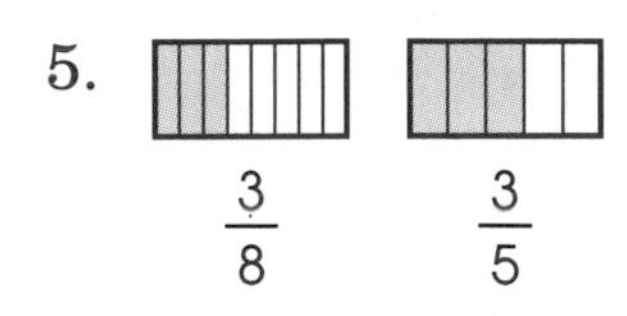

6. 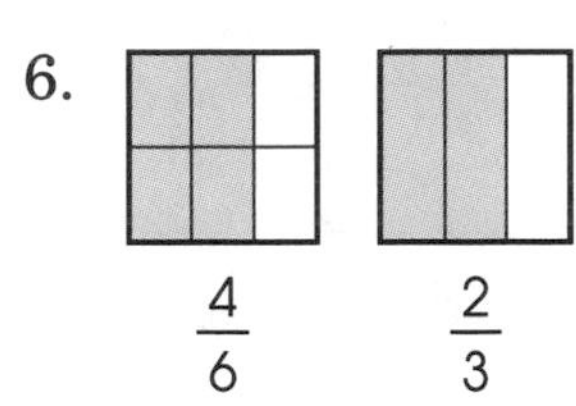

B.

You can use multiplication to find equivalent fractions. Multiply the numerator and denominator by the same number. Look at the example below. It shows that $\frac{3}{4}$ is equivalent to $\frac{6}{8}$.

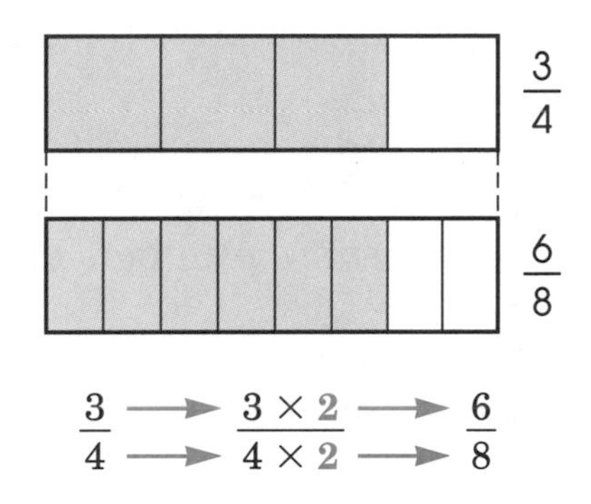

$$\frac{3}{4} \longrightarrow \frac{3 \times 2}{4 \times 2} \longrightarrow \frac{6}{8}$$

Look at the example below. Find an equivalent fraction to $\frac{1}{4}$.

Step 1 Multiply the numerator and denominator by 3. $\frac{1 \times 3}{4 \times 3} = \frac{3}{12}$

Step 2 Write the equivalent fractions. $\frac{1}{4} = \frac{3}{12}$

Copy each fraction. Multiply the numerator and denominator by 2. Write the equivalent fraction.

1. $\frac{2}{3}$
2. $\frac{4}{9}$
3. $\frac{4}{5}$
4. $\frac{5}{6}$

Copy each fraction. Multiply the numerator and denominator by 4. Write the equivalent fraction.

5. $\frac{1}{3}$
6. $\frac{3}{10}$
7. $\frac{2}{5}$
8. $\frac{1}{4}$

Lesson 22 Writing Fractions in Lowest Terms

Words to Know

reduce to divide the numerator and the denominator by the same number

lowest terms fraction with 1 as the only common factor of the numerator and denominator

Did you know $\frac{1}{2}$ cup and $\frac{2}{4}$ cup of flour are the same amount? Sometimes the fraction with the smaller numbers is easier to use.

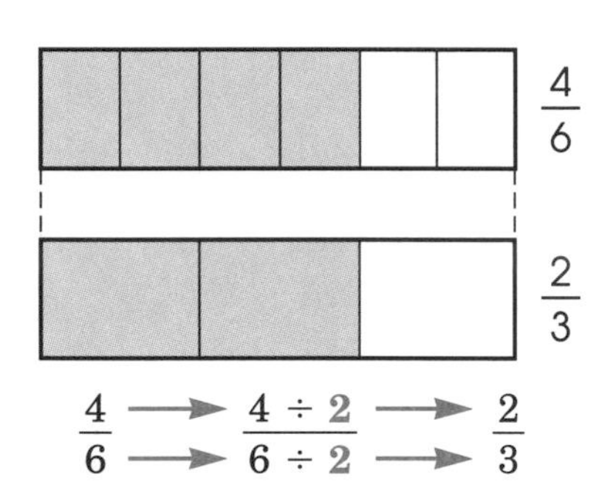

A.

You can multiply to find equivalent fractions. You can also divide to find equivalent fractions. Divide the numerator and the denominator by the same number. The art in the margin shows that $\frac{4}{6} = \frac{2}{3}$.

When you divide the numerator and denominator by the same number, you **reduce** it. Before you can divide, you need to find a common factor. Look at the example below.

Reduce. $\frac{6}{12}$

Step 1 Find a common factor of 6 and 12. 2 is a common factor. Divide by 2. $\frac{6 \div 2}{12 \div 2} = \frac{3}{6}$

Step 2 Find a common factor of 3 and 6. 3 is a common factor. Divide by 3. $\frac{3 \div 3}{6 \div 3} = \frac{1}{2}$

$\frac{6}{12}$ reduces to $\frac{1}{2}$.

Reminder

A common factor of the numerator and the denominator is a number that divides evenly into both.

Reduce each fraction by dividing the numerator and denominator by 2.

1. $\frac{2}{10}$
2. $\frac{6}{8}$
3. $\frac{4}{6}$
4. $\frac{10}{12}$

Reduce each fraction by dividing the numerator and denominator by 3.

5. $\frac{3}{12}$
6. $\frac{3}{6}$
7. $\frac{9}{15}$
8. $\frac{3}{21}$

B. A fraction is in **lowest terms** when 1 is the only common factor of the numerator and denominator. You can reduce a fraction to lowest terms by first finding the greatest common factor. Look at the example.

Find the greatest common factor of $\frac{4}{12}$.

Step 1 List the factors of the numerator and the denominator.
4: 1, 2, 4
12: 1, 2, 3, 4, 6, 12

Step 2 Find the greatest common factor of both numbers.
4: 1, 2, 4
12: 1, 2, 3, 4, 6, 12

The GCF of 4 and 12 is 4.

Reminder

GCF means greatest common factor.

List the factors. Then find the greatest common factor (GCF) for each number pair.

1. 12 and 18
2. 9 and 12
3. 21 and 27
4. 6 and 18
5. 11 and 16
6. 36 and 48

C. When you know the greatest common factor of a fraction, you can use it to reduce the fraction to lowest terms. The steps below show how to reduce $\frac{8}{10}$ to lowest terms.

Step 1 Find the greatest common factor (GCF) of the numerator and the denominator. The GCF for $\frac{8}{10}$ is 2.
8: 1, 2, 4, 8
10: 1, 2, 5, 10

Step 2 Divide both the numerator and the denominator by the GCF (2).
$\frac{8 \div 2}{10 \div 2} = \frac{4}{5}$

Step 3 The fraction $\frac{8}{10}$ reduces to $\frac{4}{5}$. It is now in lowest terms. The only number that will divide evenly into 4 and 5 is 1.

Reminder

A fraction is in lowest terms when only 1 will divide evenly into the numerator and denominator.

Reduce $\frac{8}{12}$ to its lowest terms. Show your work on your paper.

1. List the factors of 8 and 12. Find the greatest common factor.
2. Divide 8 and 12 by the greatest common factor. Write $\frac{8}{12}$ in its lowest terms.

Reduce these fractions to lowest terms. Use the steps above.

3. $\frac{4}{8}$
4. $\frac{5}{35}$
5. $\frac{6}{8}$
6. $\frac{3}{9}$
7. $\frac{6}{15}$
8. $\frac{12}{28}$
9. $\frac{6}{10}$
10. $\frac{21}{27}$

Lesson 23 Finding Common Denominators

Words to Know

like fractions fractions that have the same denominator

common denominator a common multiple of two denominators

unlike fractions fractions that have different denominators

least common denominator the smallest common denominator

All fractions do not look alike. You must look closely at each numerator and denominator.

A. **Like fractions** have the same denominator. They have a **common denominator.** The bottom numbers in both fractions are the same. For example, $\frac{5}{6}$ and $\frac{3}{6}$ are like fractions.

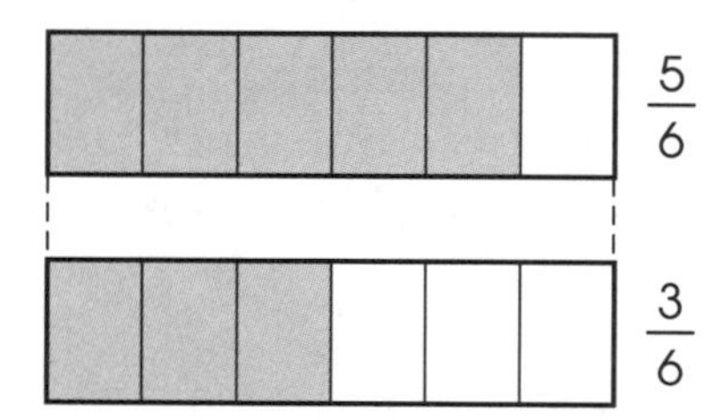

The fraction parts are the same size in like fractions. **Unlike fractions** have different denominators. $\frac{3}{4}$ and $\frac{5}{6}$ are unlike fractions. The fraction parts are different sizes.

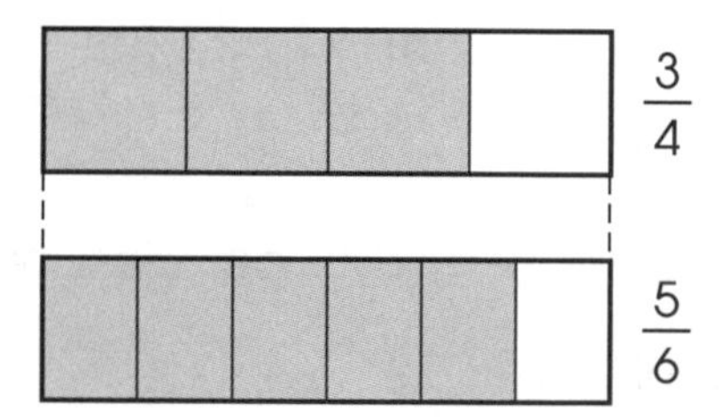

Write the fractions in each set that have common denominators.

1. $\frac{1}{3}$ $\frac{3}{4}$ $\frac{2}{3}$

2. $\frac{3}{6}$ $\frac{2}{6}$ $\frac{5}{6}$

3. $\frac{1}{5}$ $\frac{4}{5}$ $\frac{5}{10}$

4. $\frac{5}{16}$ $\frac{16}{5}$ $\frac{1}{16}$

B.

Unlike fractions have different denominators. You can change them to like fractions by finding a common denominator. This can be done in five simple steps. Look at the example below.

Change the unlike fractions $\frac{5}{6}$ and $\frac{1}{4}$ to like fractions.

Step 1 Find the least common multiple (LCM).

6: 0, 6, 12, 18
4: 0, 4, 8, 12

Step 2 The LCM is 12. Use 12 for the new denominators.

$\frac{5}{6} = \frac{\quad}{12}$ $\frac{1}{4} = \frac{\quad}{12}$

Step 3 Change $\frac{5}{6}$ to an equivalent fraction. How many times does 6 go into 12? 6×2 is 12. Multiply 5 by 2 to find the new numerator.

$\frac{5 \times 2}{6 \times 2} = \frac{10}{12}$

Step 4 Now repeat Step 3 with the next fraction $\left(\frac{1}{4}\right)$. 4×3 is 12. Multiply 1 by 3 to find the new numerator.

$\frac{1 \times 3}{4 \times 3} = \frac{3}{12}$

Step 5 Do the new fractions have the same denominators? Yes. Now $\frac{10}{12}$ and $\frac{3}{12}$ are like fractions.

Reminder

To find multiples, multiply the number by 0, 1, 2, 3, 4, 5, and so on.

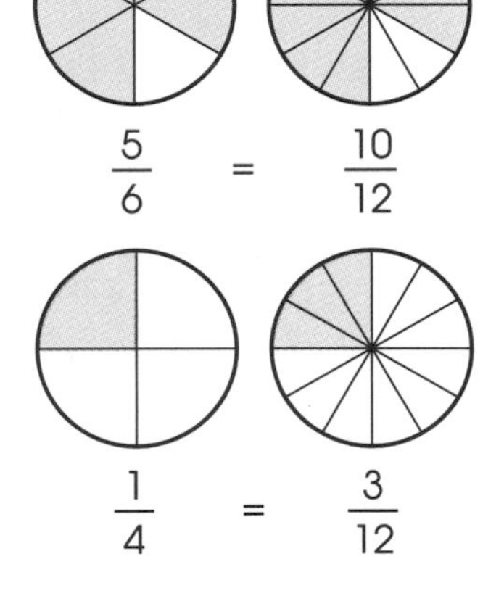

Find the common denominators of $\frac{4}{5}$ and $\frac{5}{6}$. Steps 1 and 2 have been done for you. Follow Steps 3, 4, and 5.

1. The LCM of $\frac{4}{5}$ and $\frac{5}{6}$ is 30.
2. Use 30 as the new denominator.
3. Change $\frac{4}{5}$ to an equivalent fraction.
4. Change $\frac{5}{6}$ to an equivalent fraction.
5. What are the like fractions?

C.

The **least common denominator** uses the smallest common multiple of both numbers. Always choose the least common multiple (other than zero) for the new denominator.

First, find the least common denominator for the unlike fraction pairs below. Then, change the unlike fractions to like fractions.

1. $\frac{4}{10}$ and $\frac{7}{8}$ 2. $\frac{4}{5}$ and $\frac{1}{7}$ 3. $\frac{5}{9}$ and $\frac{3}{4}$ 4. $\frac{1}{16}$ and $\frac{3}{8}$

5. $\frac{1}{5}$ and $\frac{1}{6}$ 6. $\frac{5}{6}$ and $\frac{2}{3}$ 7. $\frac{1}{4}$ and $\frac{3}{5}$ 8. $\frac{1}{5}$ and $\frac{3}{12}$

Lesson 24 Comparing Fractions

Words to Know

comparing fractions looking at fractions to decide which is larger or smaller

A friend wants to share a box of raisins with you. You have a choice between $\frac{3}{4}$ of a box or $\frac{5}{6}$ of a box. Which would you choose? To make the best choice, you need to compare the fractions. By **comparing fractions,** you can decide which fraction is larger or smaller.

A.

Comparing like fractions is not hard. They have the same denominator. Just compare the numerators, the top numbers. The fraction with the larger numerator is the larger fraction. Look at the pictures below. $\frac{5}{9}$ is larger than $\frac{3}{9}$ because 5 is larger than 3.

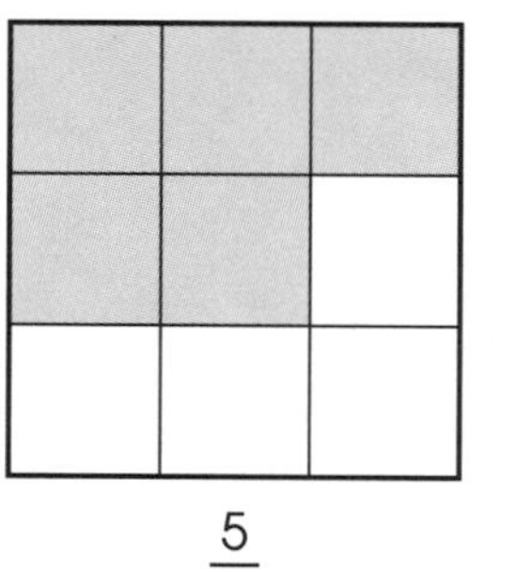

$\frac{5}{9}$

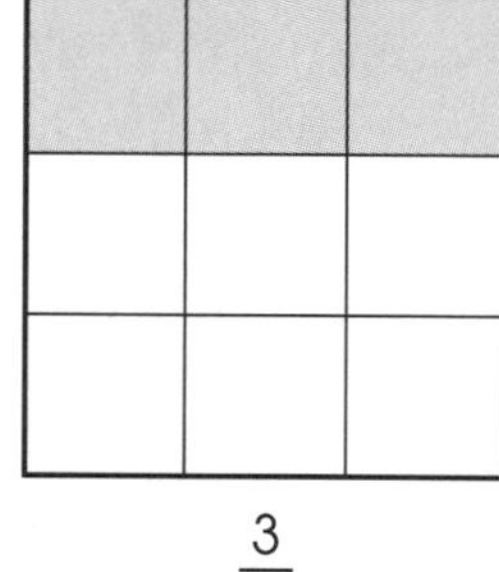

$\frac{3}{9}$

Copy the fraction pairs below. Circle the larger fraction in each.

1. $\frac{6}{8}$ and $\frac{2}{8}$
2. $\frac{11}{16}$ and $\frac{15}{16}$
3. $\frac{5}{12}$ and $\frac{6}{12}$
4. $\frac{4}{7}$ and $\frac{1}{7}$

Copy each fraction below. Write a larger fraction. Hint: Keep the same denominator and make the numerator larger.

5. $\frac{5}{8}$
6. $\frac{2}{9}$
7. $\frac{1}{5}$
8. $\frac{3}{5}$

B. You also can compare unlike fractions. Remember that unlike fractions have different denominators. Before comparing, you must change these fractions to like fractions. Like fractions have the same denominator.

Compare $\frac{3}{4}$ and $\frac{5}{6}$. Follow the steps below.

Step 1 Find the least common denominator. The least common denominator of $\frac{3}{4}$ and $\frac{5}{6}$ is 12.

4: 0, 4, 8, **12**
6: 0, 6, **12**, 18

Step 2 Change each fraction to an equivalent fraction. Remember to use 12 as the common denominator.

$$\frac{3 \times 3}{4 \times 3} = \frac{9}{12}$$
$$\frac{5 \times 2}{6 \times 2} = \frac{10}{12}$$

Step 3 Compare the numerators.

10 is larger than 9, so $\frac{5}{6}$ is larger than $\frac{3}{4}$.

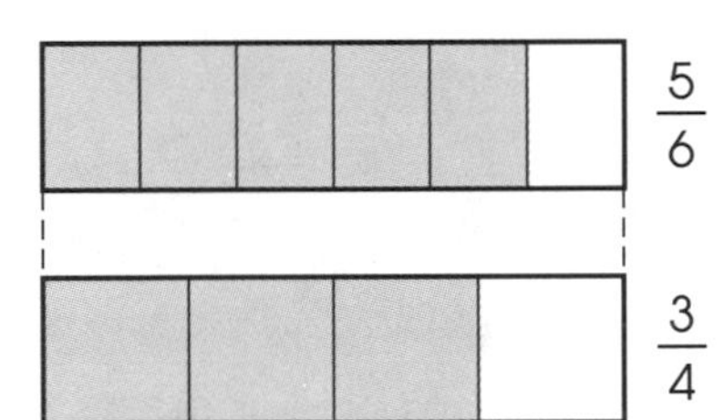

> ***Reminder***
> The common denominator is the least common multiple (LCM) of both denominators.

Compare the unlike fractions in each pair below. Change them to like fractions. Write the larger fraction.

1. $\frac{2}{3}$ and $\frac{3}{4}$
2. $\frac{1}{4}$ and $\frac{1}{5}$
3. $\frac{7}{8}$ and $\frac{5}{6}$
4. $\frac{7}{12}$ and $\frac{5}{9}$
5. $\frac{1}{2}$ and $\frac{2}{3}$
6. $\frac{3}{5}$ and $\frac{4}{7}$
7. $\frac{4}{5}$ and $\frac{7}{10}$
8. $\frac{1}{2}$ and $\frac{4}{7}$
9. $\frac{4}{5}$ and $\frac{5}{6}$
10. $\frac{3}{7}$ and $\frac{5}{8}$

> ***Reminder***
> Show your work on your paper.

Lesson 25 Changing Improper Fractions to Mixed Numbers

Words to Know

mixed number a number that is made up of a whole number and a fraction

proper fraction a fraction whose numerator is smaller than its denominator

improper fraction a fraction whose numerator is larger than its denominator

The same amounts can be expressed in many ways. An egg carton holds 12 eggs. Did you know that $1\frac{7}{12}$ cartons of eggs is the same as $\frac{19}{12}$ cartons of eggs? There are 19 eggs. Each egg is $\frac{1}{12}$ of a carton.

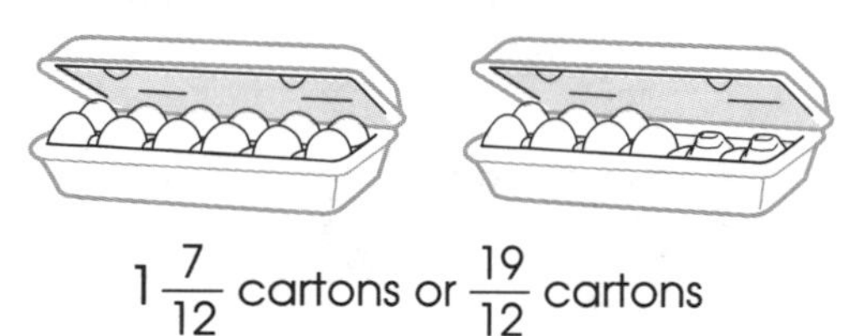

$1\frac{7}{12}$ cartons or $\frac{19}{12}$ cartons

Reminder

Whole numbers are 0, 1, 2, 3, 4, 5, and so on.

A. Look at the example above. Notice that $1\frac{7}{12}$ is a **mixed number.** A mixed number is made up of a whole number and a fraction. $\frac{19}{12}$ is not a mixed number, it is an **improper fraction.** The numerator 19 is larger than the denominator 12.

In a **proper fraction** the numerator is smaller than the denominator. Some proper fractions are:

$\frac{1}{2}$, $\frac{2}{9}$, and $\frac{13}{15}$

Copy the numbers below. Write *whole number, mixed number, proper fraction,* or *improper fraction* next to each.

1. 8
2. $\frac{4}{5}$
3. $3\frac{4}{5}$
4. $\frac{9}{4}$
5. $\frac{11}{12}$
6. $8\frac{2}{3}$
7. $\frac{9}{8}$
8. 23
9. $24\frac{5}{8}$
10. $\frac{4}{7}$

B. Improper fractions can be changed to mixed numbers. Follow the three steps below. They show how to change the improper fraction $\frac{8}{3}$ to a mixed number.

Write $\frac{8}{3}$ as a mixed number.

Step 1 Divide the numerator by the denominator.

$$3\overline{)8} = 2 \text{ R}2 \qquad 8 - 6 = 2$$

Step 2 Write the answer as a mixed number. Show the remainder as a fraction. The remainder is the numerator and the divisor is the denominator.

$$3\overline{)8} = 2\frac{2}{3}$$

remainder (2)
divisor (3)

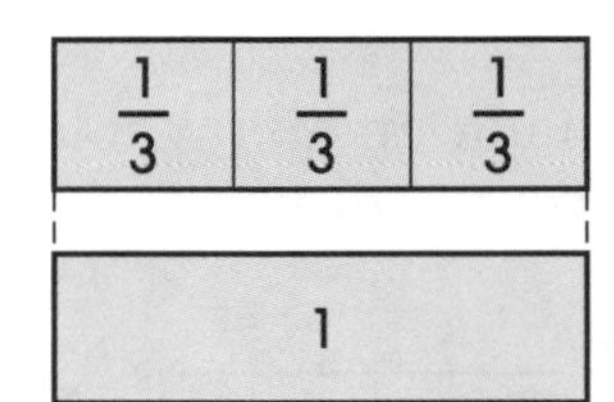

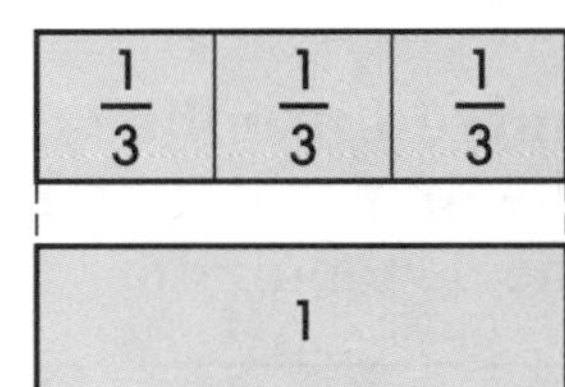

1/3 1/3 | $\frac{8}{3}$

1/3 1/3 | $2\frac{2}{3}$

The improper fraction $\frac{8}{3}$ is the same as the mixed number $2\frac{2}{3}$.

Sometimes the answer will be a whole number instead of a mixed number. This happens when there is no remainder.

Change $\frac{12}{4}$ to a whole number.

Step 1 Divide the numerator by the denominator.

$$4\overline{)12} = 3 \qquad 12 - 12 = 0$$

Step 2 The answer is 3, a whole number.

The improper fraction $\frac{12}{4}$ is the same as the whole number 3.

Change these improper fractions to mixed or whole numbers. Show your work.

1. $\frac{7}{2}$
2. $\frac{16}{4}$
3. $\frac{23}{7}$
4. $\frac{40}{5}$
5. $\frac{43}{7}$
6. $\frac{20}{6}$
7. $\frac{29}{8}$
8. $\frac{15}{2}$
9. $\frac{56}{8}$
10. $\frac{65}{10}$

Lesson 26 Changing Mixed Numbers to Improper Fractions

Words to Know

mixed number a number that is made up of a whole number and a fraction

improper fraction a fraction whose numerator is larger than its denominator

Any mixed number can be changed to an improper fraction.

A.

Remember that a **mixed number** is made up of a whole number and a fraction. The numerator of an **improper fraction** is larger than the denominator.

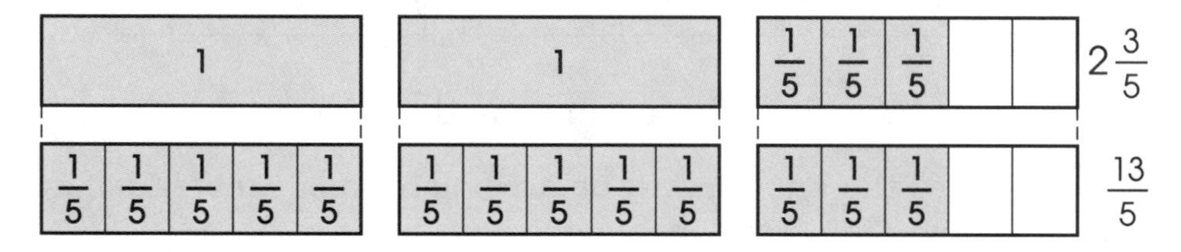

Copy the numbers below. Write *whole number, mixed number, proper fraction,* or *improper fraction* next to it.

1. $\frac{5}{2}$
2. $4\frac{1}{3}$
3. $6\frac{3}{8}$
4. 12
5. $\frac{13}{10}$
6. $\frac{16}{17}$
7. $20\frac{1}{2}$
8. $\frac{11}{5}$
9. 55
10. $\frac{25}{30}$

B.

Changing a mixed number to an improper fraction is not hard. To change $4\frac{2}{3}$ to an improper fraction, follow the steps below.

Write $4\frac{2}{3}$ as an improper fraction.

Step 1 Multiply the whole number 4 by the denominator 3. $4 \times 3 = 12$

Step 2 Add the numerator 2 to the product 12. $2 + 12 = 14$

Reminder

A product is the answer to a multiplication problem. A sum is the answer in an addition problem.

Step 3 Use this sum **14** as the numerator of the improper fraction. Keep the same denominator **3**.

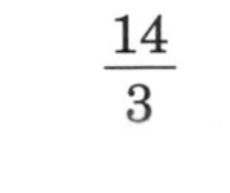

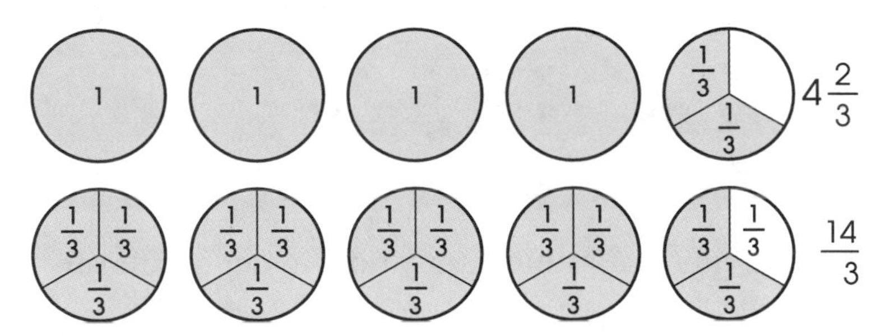

Use the steps to change $3\frac{1}{2}$ to an improper fraction.

1. Multiply the whole number by the denominator.
2. Add the numerator to the product.
3. Write the sum as the numerator of the improper fraction. Write the same denominator.

Use the steps to change $6\frac{2}{5}$ to an improper fraction.

4. Multiply the whole number by the denominator.
5. Add the numerator to the product.
6. Write the sum as the numerator of the improper fraction. Write the same denominator.

Change these mixed numbers to improper fractions. Use the steps above.

7. $4\frac{3}{5}$　8. $7\frac{3}{10}$　9. $2\frac{7}{8}$　10. $3\frac{1}{3}$

11. $4\frac{5}{7}$　12. $5\frac{4}{9}$　13. $10\frac{5}{7}$　14. $8\frac{9}{10}$

C.

Looking for mistakes is a good way to practice what you have learned.

These mixed numbers were changed to improper fractions. Copy the problems. If the work is correct, write *true*. If it is not correct, write *false*.

1. $5\frac{3}{5} = \frac{25}{5}$　2. $3\frac{7}{8} = \frac{30}{8}$　3. $4\frac{5}{6} = \frac{29}{6}$

4. $8\frac{3}{4} = \frac{33}{4}$　5. $10\frac{3}{5} = \frac{53}{5}$　6. $9\frac{4}{5} = \frac{49}{5}$

Reminder

Show your work on a separate sheet of paper.

Lesson 27 Multiplying Fractions

Word to Know

reduce to divide the numerator and the denominator by the same number

The pictures below show what $\frac{1}{2}$ of $\frac{3}{5}$ looks like.

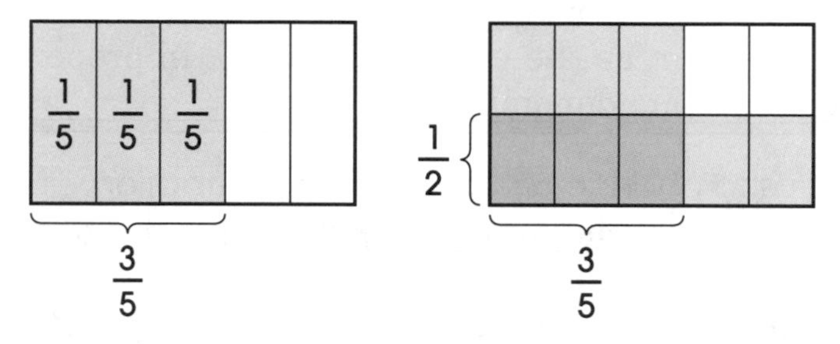

Notice $\frac{3}{10}$ of the picture is shaded dark blue. The dark blue area shows $\frac{1}{2}$ of $\frac{3}{5}$ is $\frac{3}{10}$.

A. The example above is showing multiplication. You can multiply fractions. Follow the steps to multiply $\frac{1}{2} \times \frac{3}{5}$.

Step 1 Multiply the numerators. $\frac{1}{2} \times \frac{3}{5} = \frac{3}{\ }$

Step 2 Multiply the denominators. $\frac{1}{2} \times \frac{3}{5} = \frac{3}{10}$

The product of $\frac{1}{2}$ times $\frac{3}{5}$ is $\frac{3}{10}$.

Reminder

A *product* is the answer to a multiplication problem.

Multiply the fractions below. Show your work.

1. $\frac{2}{3} \times \frac{1}{5}$
2. $\frac{3}{4} \times \frac{5}{7}$
3. $\frac{1}{2} \times \frac{3}{4}$
4. $\frac{1}{2} \times \frac{1}{2}$
5. $\frac{3}{4} \times \frac{3}{5}$
6. $\frac{3}{8} \times \frac{1}{5}$

B.

Sometimes you can **reduce** your answer. After you multiply, look for a factor of both the numerator and the denominator.

Multiply. $\frac{2}{3} \times \frac{3}{4}$

Step 1 Multiply the numerators. 2 3 6

Step 2 Multiply the denominators. 3 4 12

Step 3 Reduce the product to lowest terms.

The product of $\frac{2}{3}$ times $\frac{3}{4}$ is $\frac{6}{12}$. $\frac{6}{12}$ reduces to $\frac{1}{2}$.

Multiply. If you need to, reduce the products to lowest terms.

1. $\frac{1}{4} \times \frac{2}{5}$
2. $\frac{2}{9} \times \frac{1}{2}$
3. $\frac{2}{3} \times \frac{5}{6}$
4. $\frac{1}{2} \times \frac{4}{6}$
5. $\frac{2}{5} \times \frac{3}{8}$
6. $\frac{2}{3} \times \frac{3}{4}$
7. $\frac{5}{2} \times \frac{2}{5}$
8. $\frac{9}{10} \times \frac{5}{6}$

Reminder

A factor is a number that divides into a number with no remainder.

C.

You also can multiply fractions and whole numbers. Write the whole number as a fraction. Use 1 as the denominator.

Multiply. $\frac{1}{2} \times 5$

Step 1 Write the whole number as a fraction. $\frac{1}{2} = \frac{5}{1}$

Step 2 Multiply the numerators, and multiply the denominators. $\frac{1}{2} \times \frac{5}{1} = \frac{5}{2}$

Step 3 Write the improper fraction as a mixed number. $\frac{5}{2} = 2\frac{1}{2}$

The product of $\frac{1}{2}$ times 5 is $2\frac{1}{2}$.

Multiply. If you need to, reduce your answer to lowest terms.

1. $3 \times \frac{2}{3}$
2. $\frac{3}{4} \times 10$
3. $\frac{1}{5} \times 4$
4. $6 \times \frac{1}{2}$

Lesson 28 Multiplying Fractions with Cross Cancellation

Word to Know

cross cancellation dividing a numerator and a denominator by the same number

Sometimes there is a shortcut you can use when you multiply fractions.

The word *cancellation* sounds like kan-sehl-LAY-shuhn.

A.

Cross cancellation means dividing a numerator and denominator by the same number. You can cross-cancel only when a numerator and a denominator have a common factor. First look to see if the same number is in both a numerator and a denominator.

Multiply. $\frac{3}{4} \times \frac{1}{3}$

Step 1 Look for the same number in the numerator and denominator.

3 is in the numerator and denominator.

$\frac{3}{4} \times \frac{1}{3}$

Step 2 Divide the numerator and denominator by 3. Write the new numerator and denominator.

$\frac{\overset{1}{\cancel{3}}}{4} \times \frac{1}{\underset{1}{\cancel{3}}}$

Step 3 Multiply the new numerators and denominators.

$\frac{\overset{1}{\cancel{3}}}{4} \times \frac{1}{\underset{1}{\cancel{3}}} = \frac{1}{4}$

Reminder

Any number divided by itself is 1.

The product of $\frac{3}{4}$ times $\frac{1}{3}$ is $\frac{1}{4}$.

Multiply. Use cross cancellation first.

1. $\frac{1}{5} \times \frac{5}{6}$
2. $\frac{3}{8} \times \frac{1}{3}$
3. $\frac{1}{8} \times \frac{8}{9}$
4. $\frac{4}{5} \times \frac{3}{4}$

B.

You can use cross cancellation even when you do not see the same number in the numerator and denominator. Look for other common factors.

Multiply. $\frac{1}{2} \times \frac{4}{5}$

Step 1 Find a common factor of a numerator and denominator. $\frac{1}{2} \times \frac{4}{5}$

2 and 4 have a common factor of 2.

Step 2 Divide the denominator 2 and the numerator 4 by the common factor 2. Write the new numerator and denominator. $\frac{1}{\not{2}_{1}} \times \frac{\not{4}^{2}}{5}$

Step 3 Multiply the new numerators and denominators. $\frac{1}{\not{2}_{1}} \times \frac{\not{4}^{2}}{5} = \frac{2}{5}$

Multiply. Use cross cancellation, if possible. Show your work.

1. $\frac{3}{8} \times \frac{1}{6}$
2. $\frac{3}{4} \times \frac{6}{7}$
3. $\frac{3}{5} \times \frac{5}{8}$
4. $\frac{3}{4} \times \frac{5}{9}$
5. $\frac{2}{5} \times \frac{3}{10}$
6. $\frac{2}{15} \times \frac{10}{11}$
7. $\frac{1}{9} \times \frac{3}{4}$
8. $\frac{4}{5} \times \frac{7}{12}$

C.

Sometimes you can cross-cancel more than once. You can cancel two times in the example below.

Multiply. $\frac{3}{10} \times \frac{6}{15}$

Step 1 Find common factors of the numerators and denominators. $\frac{3}{10} \times \frac{6}{15}$

3 and 15 have a common factor of 3. 6 and 10 have a common factor of 2.

Step 2 Divide the numerators and denominators by the common factors. Write the new numbers. $\frac{\not{3}^{1}}{\not{10}_{5}} \times \frac{\not{6}^{3}}{\not{15}_{5}}$

Step 3 Multiply the new numerators and denominators. $\frac{\not{3}^{1}}{\not{10}_{5}} \times \frac{\not{6}^{3}}{\not{15}_{5}} = \frac{3}{25}$

Multiply. Use cross cancellation, if you can.

1. $\frac{3}{4} \times \frac{8}{15}$
2. $\frac{5}{6} \times \frac{2}{10}$
3. $\frac{8}{15} \times \frac{5}{6}$
4. $\frac{5}{12} \times \frac{3}{5}$

Lesson 29 Dividing Fractions

Word to Know

invert to reverse, or switch, the positions of the numerator and denominator in a fraction

Sometimes you may need to divide fractions. How many $\frac{1}{8}$ foot pieces can you cut from a board that is $\frac{1}{2}$ foot long? You have to find out how many $\frac{1}{8}$s fit into $\frac{1}{2}$.

Divide $\frac{1}{2}$ by $\frac{1}{8}$.

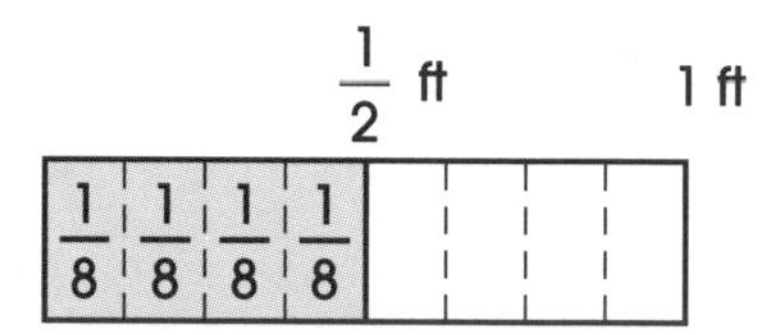

There are four $\frac{1}{8}$s in $\frac{1}{2}$. Dividing by $\frac{1}{8}$ is the same as multiplying by 8.

$$\frac{1}{2} \div \frac{1}{8} = 4$$

$$\frac{1}{2} \times \frac{8}{1} = \frac{1 \times 8}{2 \times 1} = \frac{8}{2} = \frac{4}{1} = 4.$$

A. The first step in dividing fractions is to **invert** the divisor. How do you invert a fraction? You reverse, or switch, the positions of the numerator and denominator. You can invert a whole number, too. Write the whole number as a fraction, then invert.

$\frac{5}{8}$ inverted is $\frac{8}{5}$ 3 inverted is $\frac{1}{3}$

Copy the numbers. Invert the numbers.

1. $\frac{2}{3}$ 2. $\frac{7}{4}$ 3. $\frac{1}{6}$ 4. $\frac{5}{9}$ 5. 4

B.

Look at the steps below. The three steps show you how to divide a fraction by another fraction.

Divide. $\frac{2}{3} \div \frac{4}{5}$

Step 1 Invert the divisor. $\frac{5}{4}$

Step 2 Change the division sign to a multiplication sign. $\frac{2}{3} \times \frac{5}{4}$

Step 3 Cross cancel if you can. Multiply the numerators. Multiply the denominators. $\frac{\cancel{2}^{1}}{3} \times \frac{5}{\cancel{4}_{2}} = \frac{1 \times 5}{3 \times 2} = \frac{5}{6}$

The quotient of $\frac{2}{3}$ divided by $\frac{4}{5}$ is $\frac{5}{6}$.

> **Reminder**
>
> dividend ÷ divisor = quotient. The divisor is the number to divide by. The divisor in $\frac{3}{10} \div \frac{1}{5}$ is $\frac{1}{5}$.

Divide. Show your work.

1. $\frac{1}{2} \div \frac{5}{12}$
2. $\frac{5}{7} \div \frac{1}{2}$
3. $\frac{3}{4} \div \frac{3}{4}$
4. $\frac{5}{9} \div \frac{2}{3}$
5. $\frac{3}{8} \div \frac{1}{6}$
6. $\frac{4}{9} \div \frac{2}{3}$
7. $\frac{1}{2} \div \frac{1}{4}$
8. $\frac{3}{4} \div \frac{1}{8}$
9. $\frac{3}{5} \div \frac{1}{5}$

C.

You also can divide fractions and whole numbers. Write the whole number as a fraction. Use 1 as the denominator.

Divide. $3 \div \frac{2}{3}$

Step 1 Write the whole number as a fraction. $\frac{3}{1} \div \frac{2}{3}$

Step 2 Invert the divisor. Change the division sign to a multiplication sign. $\frac{3}{1} \times \frac{3}{2}$

Step 3 Multiply the numerators and denominators. $\frac{3 \times 3}{1 \times 2} = \frac{9}{2}$

Step 4 Write the improper fraction as a mixed number. $\frac{9}{2} = 4\frac{1}{2}$

The quotient of 3 divided by $\frac{2}{3}$ is $4\frac{1}{2}$.

> **Reminder**
>
> To change an improper fraction to a mixed number, divide the numerator by the denominator. Put the remainder over the denominator.

Divide. Show your work.

1. $\frac{1}{2} \div 3$
2. $4 \div \frac{2}{3}$
3. $1 \div \frac{1}{4}$
4. $\frac{2}{3} \div 1$
5. $6 \div \frac{1}{3}$
6. $\frac{3}{4} \div 8$
7. $1 \div \frac{3}{7}$
8. $5 \div \frac{3}{4}$

Lesson 30 Adding and Subtracting Like Fractions

Words to Know

like fractions fractions that have the same denominator

Two students worked together on a science project. Alan did $\frac{3}{10}$ of the project. Chris did $\frac{6}{10}$ of the project. How much of the project is finished? Add the fractions $\frac{3}{10}$ and $\frac{6}{10}$. $\frac{9}{10}$ of the project is finished.

A.

Like fractions have the same denominator. Adding like fractions is easy. Follow the steps below.

$\frac{1}{8}$	$\frac{1}{8}$	$\frac{1}{8}$	$\frac{1}{8}$	$\frac{1}{8}$			

$\frac{2}{8}$ $\frac{3}{8}$

$\frac{2}{8} + \frac{3}{8} = \frac{5}{8}$

Add. $\frac{2}{8} + \frac{3}{8}$

Step 1 Add the numerators.

Step 2 Keep the same denominator. $\frac{2}{8} + \frac{3}{8} = \frac{5}{8}$ or $\begin{array}{r} \frac{2}{8} \\ + \frac{3}{8} \\ \hline \frac{5}{8} \end{array}$

Add. Show your work.

1. $\frac{3}{10} + \frac{4}{10}$
2. $\frac{3}{7} + \frac{2}{7}$
3. $\frac{1}{9} + \frac{7}{9}$
4. $\frac{3}{5} + \frac{1}{5}$
5. $\frac{1}{4} + \frac{2}{4}$
6. $\frac{2}{11} + \frac{3}{11}$
7. $\frac{1}{5} + \frac{1}{5}$
8. $\frac{4}{9} + \frac{1}{9}$

B.

Sometimes you will have to reduce your answer to lowest terms. Look at your answer. Is there a common factor in the numerator and denominator? If so, reduce to lowest terms.

Add. $\frac{3}{10} + \frac{2}{10}$

Step 1 Add the numerators. Keep the denominator the same. $\frac{3}{10} + \frac{2}{10} = \frac{5}{10}$ or $\begin{array}{r} \frac{3}{10} \\ + \frac{2}{10} \\ \hline \frac{5}{10} \end{array}$

Step 2 Write the fraction in lowest terms. 5 is the greatest common factor of 5 and 10. $\frac{5 \div 5}{10 \div 5} = \frac{1}{2}$

The sum of $\frac{3}{10} + \frac{2}{10}$ is $\frac{5}{10}$ which reduces to $\frac{1}{2}$.

Add. Reduce the sums to lowest terms.

1. $\frac{1}{8} + \frac{3}{8}$ 2. $\frac{1}{4} + \frac{1}{4}$ 3. $\frac{5}{12} + \frac{3}{12}$ 4. $\frac{3}{10} + \frac{5}{10}$

Reminder

To reduce a fraction to lowest terms, divide the numerator and the denominator by their *greatest common factor.*

C.

Follow the steps below to subtract like fractions.

Subtract. $\frac{4}{5} - \frac{2}{5}$

Step 1 Subtract the numerators.

Step 2 Keep the same denominator. $\frac{4}{5} - \frac{2}{5} = \frac{2}{5}$ or $\begin{array}{r} \frac{4}{5} \\ -\frac{2}{5} \\ \hline \frac{2}{5} \end{array}$

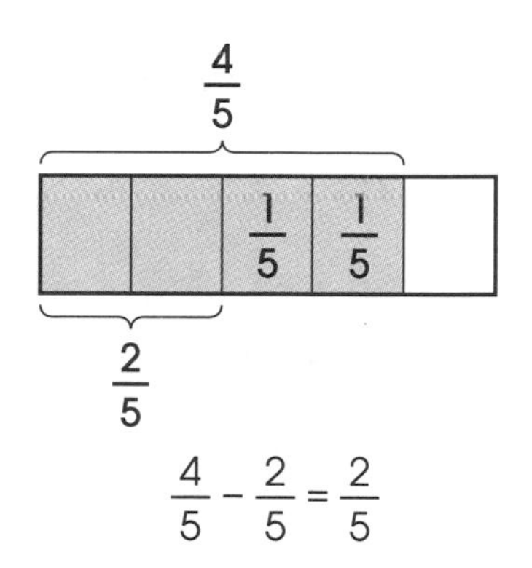

Subtract. Reduce each answer to lowest terms.

1. $\frac{8}{9} - \frac{3}{9}$ 2. $\frac{8}{8} - \frac{3}{8}$ 3. $\frac{5}{7} - \frac{2}{7}$ 4. $\frac{4}{6} - \frac{1}{6}$

5. $\frac{7}{10} - \frac{3}{10}$ 6. $\frac{5}{9} - \frac{2}{9}$ 7. $\frac{11}{12} - \frac{7}{12}$ 8. $\frac{5}{6} - \frac{1}{6}$

D.

You also can subtract a fraction from the whole number 1. Write 1 as a fraction using the denominator of the other fraction. Follow the steps below.

Subtract. $1 - \frac{2}{3}$

Step 1 Write 1 as a fraction. Use 3 as the denominator. $1 = \frac{3}{3}$

Step 2 Subtract the numerators. Keep the denominator the same. $\frac{3}{3} - \frac{2}{3} = \frac{1}{3}$

The difference between 1 and $\frac{2}{3}$ is $\frac{1}{3}$.

Subtract. Reduce each answer to lowest terms.

1. $1 - \frac{3}{4}$ 2. $1 - \frac{4}{10}$ 3. $1 - \frac{5}{7}$ 4. $1 - \frac{2}{5}$

Lesson 31 Adding and Subtracting Unlike Fractions

Words to Know

unlike fractions fractions that have different denominators

Mrs. Jackson is baking cookies. She needs to add $\frac{1}{2}$ cup flour and $\frac{1}{3}$ cup flour. How much flour does she need all together? These two measurements are unlike fractions. This lesson will show you how to add unlike fractions.

A.

Unlike fractions have different denominators. To add these fractions, first you need to change the *unlike* fractions to *like* fractions. Then you can add. Look at the steps below.

$\frac{1}{2}$	$\frac{1}{3}$	

$\frac{1}{6}$	$\frac{1}{6}$	$\frac{1}{6}$	$\frac{1}{6}$	$\frac{1}{6}$	

$\frac{3}{6}$ $\frac{2}{6}$

$\frac{1}{2} + \frac{1}{3}$

Add. $\frac{1}{2} + \frac{1}{3}$

$\frac{1}{2}$	$\frac{1}{3}$	

Step 1 Find the least common denominator. Write the equivalent fractions.

$$\frac{1 \times 3}{2 \times 3} = \frac{3}{6} \qquad \frac{1 \times 2}{3 \times 2} = \frac{2}{6}$$

Step 2 Add the numerators. Keep the same denominator.

$$\frac{3}{6} + \frac{2}{6} = \frac{5}{6} \quad \text{or} \quad \begin{array}{r} \frac{1}{2} = \frac{3}{6} \\ + \frac{1}{3} = \frac{2}{6} \\ \hline \frac{5}{6} \end{array}$$

The sum of $\frac{1}{2}$ and $\frac{1}{3}$ is $\frac{5}{6}$.

Add. Reduce the sums to lowest terms.

1. $\frac{2}{5} + \frac{1}{2}$
2. $\frac{1}{4} + \frac{2}{3}$
3. $\frac{2}{3} + \frac{1}{12}$
4. $\frac{1}{4} + \frac{3}{8}$
5. $\frac{1}{3} + \frac{1}{6}$
6. $\frac{2}{6} + \frac{5}{12}$
7. $\frac{2}{3} + \frac{1}{5}$
8. $\frac{1}{6} + \frac{2}{9}$

B.

Sometimes when you add fractions you will get an improper fraction. Write the improper fraction as a mixed number.

> **Reminder**
>
> To find the least common denominator, list the multiples of the denominators. The smallest multiple they share is the least common denominator.

Add. $\frac{3}{4} + \frac{1}{2}$

Step 1 Find the least common denominator. Write the equivalent fractions. $\frac{3}{4} = \frac{3}{4} \quad \frac{1}{2} = \frac{1 \times 2}{2 \times 2} = \frac{2}{4}$

Step 2 Add the numerators. Keep the denominator the same. $\frac{3}{4} + \frac{2}{4} = \frac{5}{4}$

Step 3 Write the improper fraction as a mixed number. $\frac{5}{4} = 1\frac{1}{4}$

The sum of $\frac{3}{4}$ and $\frac{1}{2}$ is $1\frac{1}{4}$.

Add. Write improper fractions as mixed numbers.

1. $\frac{2}{3} + \frac{3}{4}$
2. $\frac{3}{5} + \frac{7}{10}$
3. $\frac{1}{3} + \frac{5}{6}$
4. $\frac{1}{2} + \frac{4}{5}$

C.

You cannot subtract unlike fractions. First you must change them to like fractions. Look at the steps below.

Subtract. $\frac{2}{3} - \frac{1}{2}$

Step 1 Find the least common denominator. Write the equivalent fractions. $\frac{2 \times 2}{3 \times 2} = \frac{4}{6}$ $\frac{1 \times 3}{2 \times 3} = \frac{3}{6}$

Step 2 Subtract the numerators. Keep the same denominator. $\frac{4}{6} + \frac{3}{6} = \frac{1}{6}$ or

$$\begin{array}{rl} \frac{2}{3} = & \frac{4}{6} \\ -\frac{1}{2} = & \frac{3}{6} \\ \hline & \frac{1}{6} \end{array}$$

The difference between $\frac{2}{3}$ and $\frac{1}{2}$ is $\frac{1}{6}$.

Subtract. Reduce the fraction to lowest terms if needed.

1. $\frac{3}{4} - \frac{1}{6}$
2. $\frac{2}{3} - \frac{1}{5}$
3. $\frac{1}{4} - \frac{1}{8}$
4. $\frac{4}{5} - \frac{1}{2}$
5. $\frac{5}{6} - \frac{1}{3}$
6. $\frac{1}{2} - \frac{1}{6}$
7. $\frac{3}{4} - \frac{5}{12}$
8. $\frac{2}{3} - \frac{2}{4}$

Lesson 32 Fractions: Problem Solving

Words to Know

addition clue words words or groups of words that tell you to add

subtraction clue words words or groups of words that tell you to subtract

multiplication clue words words or groups of words that tell you to multiply

division clue words words or groups of words that tell you to divide

Clue words are important. They can help you decide how to solve a math problem. Read the clue words below.

addition clue words	in all, together, altogether, total, both
subtraction clue words	how many more, how many fewer, less how much more, remain, left, difference, greater
multiplication clue words	of, part, at, for, in, per
division clue words	into, how many, how many times, how many did each

A. Addition, subtraction, multiplication, and division are called arithmetic operations. You can use these operations to solve problems with fractions.

Reminder

Not all problems have clue words.

Copy each problem below. Circle the clue words. Write which operation you would use.

1. Diane jogged $\frac{3}{4}$ mile for 3 days in a row. How far did she jog?
2. Andy painted $\frac{3}{10}$ of the room on Monday. Jan painted $\frac{5}{10}$ on Tuesday. Altogether, how much of the room was painted?
3. Sue bought 9 yards of fabric. If one blouse uses $\frac{7}{8}$ of a yard of fabric, how many blouses can she make?

B. Six simple steps can help you solve math problems. Read the problem and the steps. Notice how the steps work.

The bookshelf is 45 inches long. Each book is $\frac{3}{4}$ inch wide. How many books will fit on the shelf?

Step 1	Read the problem.	
Step 2	Decide what you must find out.	How many books will fit on the shelf?
Step 3	Notice the clue words. Decide which operation to use.	How many tells you to divide.
Step 4	Write down the problem.	$45 \div \frac{3}{4}$
Step 5	Solve the problem. Cross-cancel if you can. Multiply the numerators and denominators.	$\frac{\overset{15}{\cancel{45}}}{1} \times \frac{4}{\underset{1}{\cancel{3}}}$ = **60 books**
Step 6	Check the answer. Does it makes sense?	

Follow the steps and answer the questions. Solve the problems.

1. On Saturday, $\frac{3}{10}$ inch of rain fell. On Sunday, $\frac{3}{4}$ inch of rain fell. How much rain fell on both days?
   a. What are the clue words?
   b. Which operation will you use?
   c. Solve the problem.

2. A total of 120 people are waiting to buy concert tickets There are enough tickets for only $\frac{3}{4}$ of them. How many of the people will get tickets?
   a. What are the clue words?
   b. Which operation will you use?
   c. Solve the problem.

3. Bill has $\frac{3}{4}$ pound of peanuts. He shares $\frac{2}{3}$ pound with his friends. How many pounds does he have left?
   a. What are the clue words?
   b. Which operation will you use?
   c. Solve the problem.

Reminder

Copy the problems and show your work.

Unit 3 Decimals

Lesson 33 What Is a Decimal?

Words to Know

decimal a number written with a dot followed by places to the right; the digits to the right of the dot stand for less than one whole

decimal point the dot in a decimal number

mixed decimal a number containing a whole number and a decimal number

The word *decimal* sounds like DEHS-uh-muhl.

Every time you walk into a store you see decimals. Prices are written in decimals. A soft drink costs \$.56, or a sweater costs \$29.98.

A. Like a fraction, a **decimal** names a part of a whole. But the whole is always 10, 100, or 1,000. Also, decimals and fractions are written in different ways. A decimal is written with a dot followed by digits to the right of the dot. The dot is called a **decimal point.** The numbers below are decimals:

.6 .08 .47 .391 .50

Write the decimal in each group of three numbers.

1) $\frac{1}{4}$ .04 4 2) $16\frac{1}{2}$ .016 $\frac{16}{17}$ 3) 8 $\frac{1}{8}$ .80

B. A place value chart tells you the value of the digits to the right of the decimal. Look at the numbers in the place value chart in the margin.

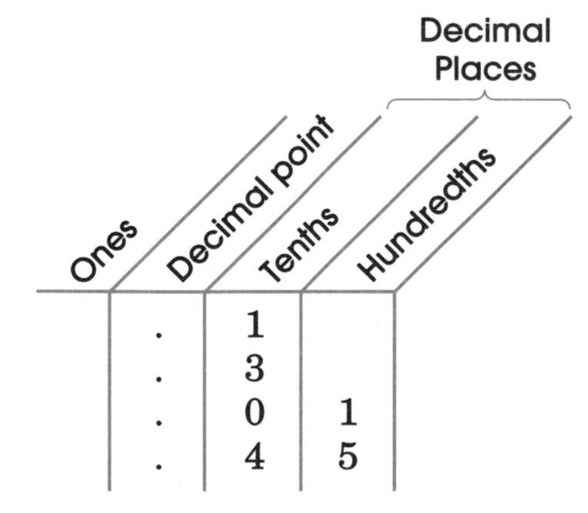

Ones	Decimal point	Tenths	Hundredths
	.	1	
	.	3	
	.	0	1
	.	4	5

.1 is read as "one tenth."
As a fraction it's $\frac{1}{10}$.

.3 is read as "three tenths."
As a fraction it's $\frac{3}{10}$.

.01 is read as "one hundredth."
As a fraction it's $\frac{1}{100}$.

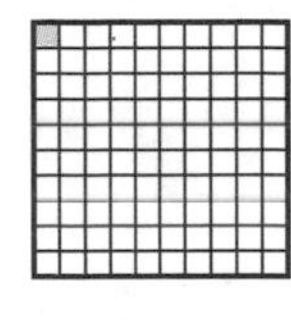

.45 is read as "forty-five hundredths."
As a fraction it's $\frac{45}{100}$.

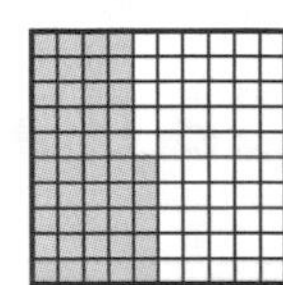

Write a decimal number for each picture below.

1) 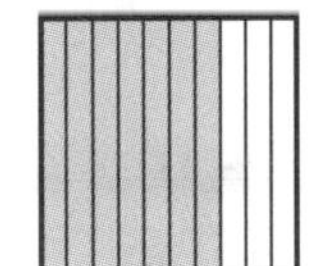
2) 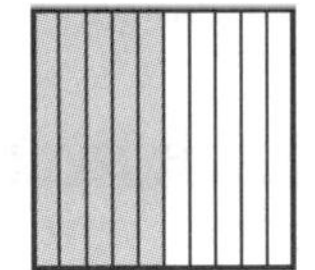
3) 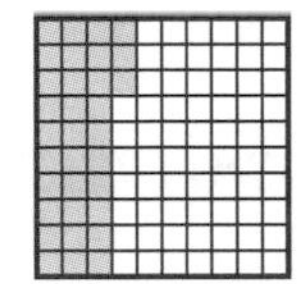
4) 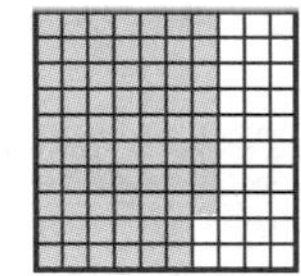

Write these decimals in words.

1) .4 2) .07 3) .8 4) .36

C.

You can write a decimal with a whole number. This is called a **mixed decimal**. The digits to the *left* of the decimal point stand for a whole number. The digits to the *right* of the decimal point stand for *less than* a whole number. Look at the picture and place value chart for 2.25.

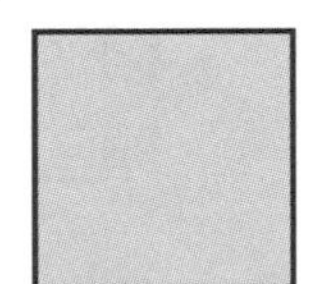
.

2 wholes and 25 hundredths

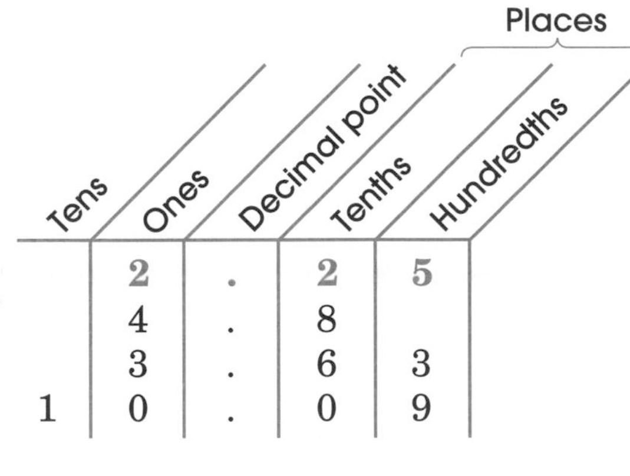

Tens	Ones	Decimal point	Tenths	Hundredths
	2	.	2	5
	4	.	8	
	3	.	6	3
1	0	.	0	9

Decimal Places: Tenths, Hundredths

2.25 is read as "two and twenty-five hundredths."

How do you read the other mixed decimals on the chart?

4.8 ⟶ four and eight tenths
3.63 ⟶ three and sixty-three hundredths
10.09 ⟶ ten and nine hundredths

Write the value of the digit 5 in each number below. Use a place value chart to help you.

1) 1.05 2) 4.52 3) .65 4) 5.37

Copy the decimal numbers. Write the decimal in words next to it.

5) 2.3 6) 6.04 7) .35 8) 3.75

9) 25.6 10) .12 11) 1.9 12) 30.02

Reminder

Read and write the decimal point as "and."

Lesson 34 Comparing Decimals

Words to Know

comparing decimals finding which decimal is larger or smaller

Tracy lives .6 miles from school. Anita lives .25 miles from school. Who lives farther from school? You need to compare the decimals to find out. Tracy lives farther.

A. **Comparing decimals** lets you find which decimal is larger or smaller. Look at the pictures below. Notice the part of each picture that is shaded.

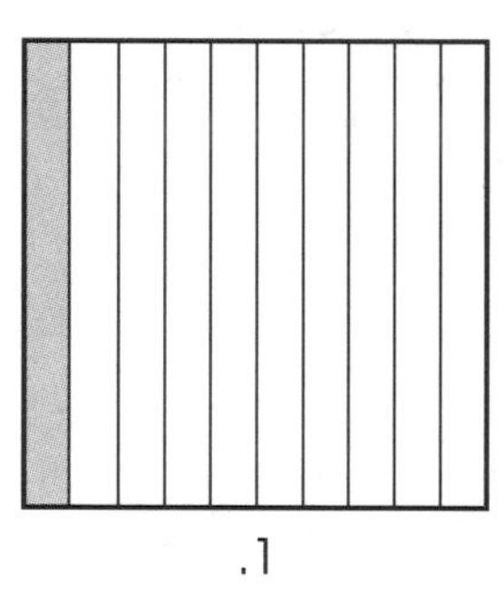

.1

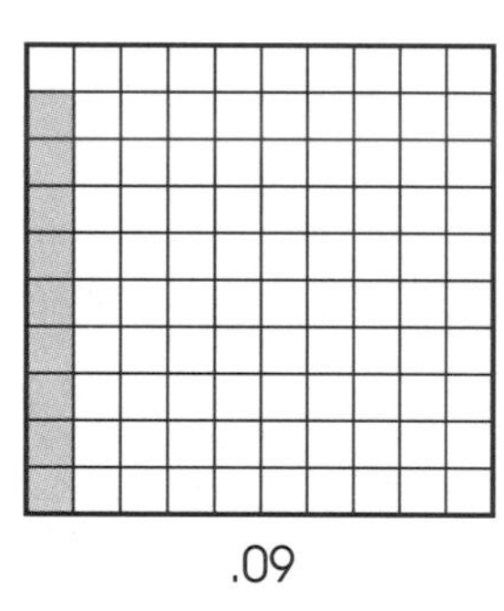

.09

You can see that .1 is larger than .09.

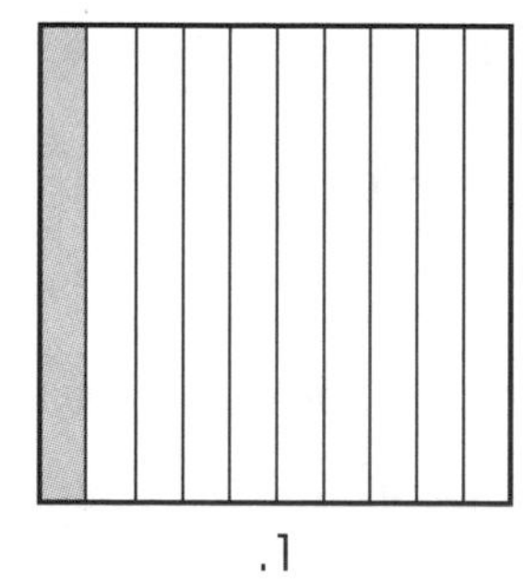

.1

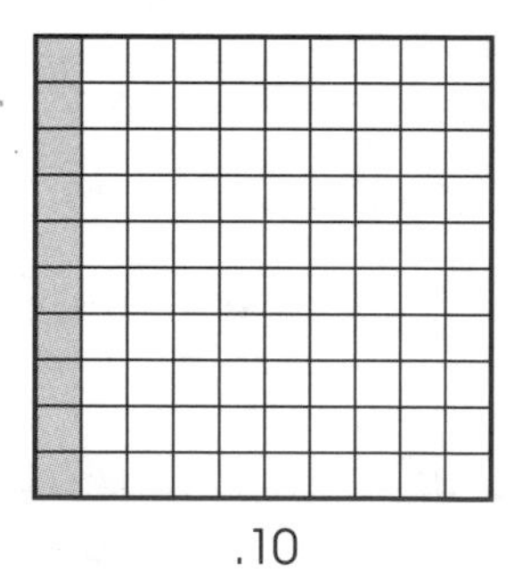

.10

.1 is the same as .10.

Look at these pictures of decimals.

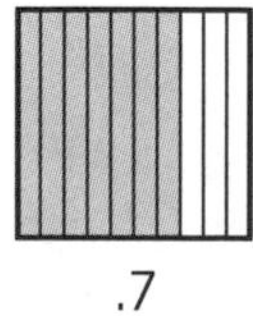

.7

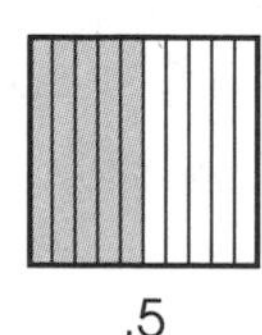

.5

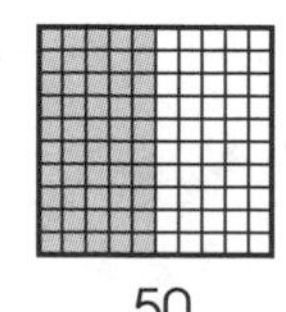

.50

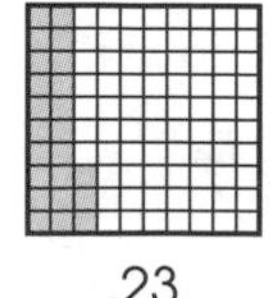

.23

.5 is smaller than **.7**.

.5 is the same as **.50**.

.5 is larger than **.23**.

Write the decimal for each picture below. Circle the larger decimal in each pair.

1) 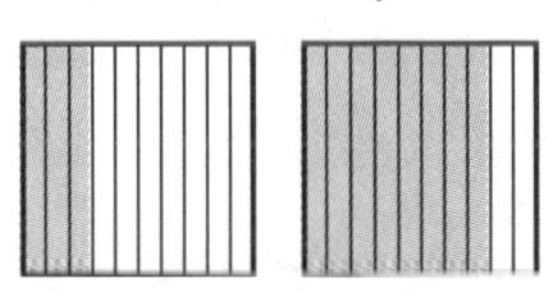

2) 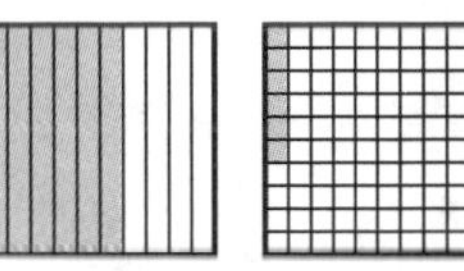 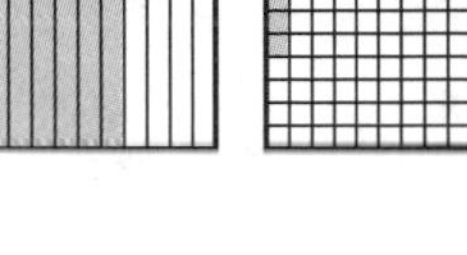

3) 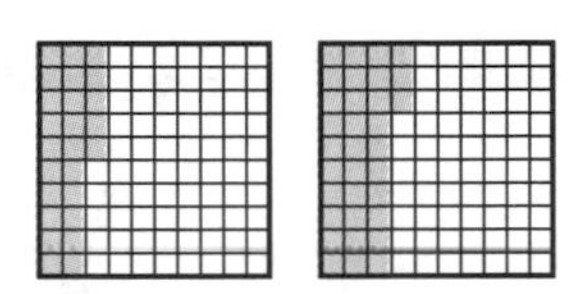

4) 

> ***Reminder***
>
> Show your work on a separate sheet of paper.

B.

You can compare decimals without looking at pictures. The three steps below show you how to compare two decimals. Look at the steps.

Compare 25.62 and 25.68.

Step 1 Line up the decimal points.

25.62
25.68

Step 2 Start at the left. Compare digits in the same places.

2, **5**, and **6** are the same.

25.62
25.68

Step 3 The first digit that is larger tells the larger decimal.

You can see that **8** is larger than **2**.

That means 25.68 is larger than 25.62.

25.6**2**
25.6**8**

Copy and compare the decimal pairs. Circle the larger decimal in each.

1) .5 .6 2) .4 .04 3) .6 .62 4) .70 .69
5) .703 .73 6) .63 .619 7) .13 .31 8) .09 .01

You can follow the same steps to compare three or more decimals.

Write the decimals in order from smallest to largest.

9) .003 .3 .05 10) .6 .61 .56
11) .43 .49 .4 12) .82 .830 .84

Lesson 35 Adding and Subtracting Decimals

Words to Know

total the answer to an addition problem
decimal point the dot in a decimal number

When you add up prices to get a **total,** you are adding decimals. Ron went to a take-out restaurant. He ordered a hamburger for $2.99, a soft drink for $.99, and french fries for $1.19. To find the total, he had to add $2.99 + $.99 + $1.19. He spent a total of $5.17.

A. You add decimals the same way you add whole numbers. To add decimals, you must line up the **decimal points.** Look at the four steps below.

Add. $42.67 + $12 + $.85

Step 1 Line up the decimal points as you write the problem.

Step 2 Use 0s to hold places.

```
  $42.67
   12.00
+    .85
```

Step 3 Add the decimals just the way you add whole numbers.

Step 4 Put a decimal point in the answer. If you are adding money, write a dollar sign.

```
  $42.67
   12.00
+    .85
  $55.52
```

Add. Remember to place the decimal point in the answer. Write a dollar sign if needed.

1) 3.4 + 12.65 + 6

2) $30 + $3.78 + $.52

3)
```
  0.6
+ 0.5
```

4)
```
  12.6
+  6.5
```

5)
```
  $ 9.24
     .28
+  54.20
```

6)
```
  1.3
+ 8.7
```

7)
```
  15.36
+   .46
```

8)
```
  13.75
    .50
+  5.25
```

B. You can subtract decimals the same way you subtract whole numbers. Look at the four steps below.

Subtract. 81.62 – 12.4

Step 1 Line up the decimal points as you write the problem.

$$\begin{array}{r} 81.62 \\ -\ 12.40 \\ \hline \end{array}$$

Step 2 Use 0s to hold places.

Step 3 Subtract the decimals just the way you subtract whole numbers.

$$\begin{array}{r} {}^{7}\not{8}{}^{1}1.62 \\ -\ \not{1}2.40 \\ \hline 69.22 \end{array}$$

Step 4 Put a decimal point in the answer. Write a dollar sign if you are adding money.

Subtract. Remember to show the decimal point in the answer. Write a dollar sign if needed.

1) 9.76 – 5.3

2) \$12.45 – \$7.32

3) \$27 – \$5.23

4) $\begin{array}{r} 9.7 \\ -\ 3.4 \\ \hline \end{array}$

5) $\begin{array}{r} \$35.39 \\ -\ 14.04 \\ \hline \end{array}$

6) $\begin{array}{r} 58.67 \\ -\ 36.04 \\ \hline \end{array}$

7) $\begin{array}{r} 10.00 \\ -\ 4.85 \\ \hline \end{array}$

8) $\begin{array}{r} 16.95 \\ -\ 7.00 \\ \hline \end{array}$

9) $\begin{array}{r} \$67.78 \\ -\ 14.23 \\ \hline \end{array}$

Practice adding and subtracting decimals. Show your work on a separate paper.

10) 18.56 – 6.15

11) 12.5 – 3.21

12) \$10 – 4.75

13) $\begin{array}{r} 13.60 \\ -\ 4.2 \\ \hline \end{array}$

14) $\begin{array}{r} 17.41 \\ +\ 3.02 \\ \hline \end{array}$

15) $\begin{array}{r} \$26.29 \\ +\ 3.70 \\ \hline \end{array}$

16) $\begin{array}{r} 16.30 \\ -\ 10.27 \\ \hline \end{array}$

17) $\begin{array}{r} 41.03 \\ -\ 5.97 \\ \hline \end{array}$

18) $\begin{array}{r} 14.32 \\ 6.7 \\ +\ 12.04 \\ \hline \end{array}$

Lesson 36 Multiplying Decimals

Words to Know

decimal places the places to the right of the decimal point—tenths, hundredths, thousandths

Terri bought three T-shirts at $12.99 each. How do you find the total Terri spent? You can multiply $12.99 by 3. Terri spent $38.97.

A. You multiply decimals the same way you multiply whole numbers. But how do you decide where to put the decimal point in the answer? You count the **decimal places.** Look at the three steps below.

Multiply. 1.3×25.4

Step 1 Multiply just as you do with whole numbers.

$$\begin{array}{r} 25.4 \\ \times \quad 1.3 \\ \hline 762 \\ 2540 \\ \hline 3302 \end{array}$$

Step 2 Count the total numer of decimal places in the numbers you have multiplied. There are two.

Step 3 Look at the product. Start at the right. Count over two places. Put in a decimal point.

$$\begin{array}{rl} 25.4 & \leftarrow \text{1 decimal place} \\ \times \quad 1.3 & \leftarrow \text{1 decimal place} \\ \hline 762 & \\ 2540 & \\ \hline 33.02 & \leftarrow \text{2 decimal places} \end{array}$$

Here are two more examples.

$$\begin{array}{rl} 5.3 & \leftarrow \text{1 decimal place} \\ \times \quad 7 & \leftarrow \text{0 decimal places} \\ \hline 37.1 & \leftarrow \text{1 decimal place} \end{array}$$

$$\begin{array}{rl} 7.62 & \leftarrow \text{2 decimal places} \\ \times \quad 2.9 & \leftarrow \text{1 decimal place} \\ \hline 6858 & \\ 15240 & \\ \hline 22.098 & \leftarrow \text{3 decimal places} \end{array}$$

Multiply. Place the decimal point in each product.

1) $.3 \times 7$

2) $.52 \times .6$

3) $37 \times .5$

4) 2.3×4.5

5) $31.2 \times .8$

6) 4.6×1.4

7) 3.48×4

8) $23.5 \times .6$

9) $.45 \times .03$

Reminder

The product is the result of a multiplication problem.

B.

You can use a quick way to multiply decimals by 10, 100, or 1,000. Look at the two steps below.

Multiply. 45.62×10
45.62×100
$45.62 \times 1{,}000$

Step 1 Count the 0s in the 10, 100, or 1,000.

Step 2 Move the decimal point one place to the right for each 0. Fill in with extra 0s if you need to.

There is one 0 in 10. Move the decimal point one place to the right. $45.62 \times 10 = 456.2 = 456.2$

There are two 0s in 100. Move the decimal point two places to the right. $45.62 \times 100 = 4562 = 4{,}562$

There are three 0s in 1,000. Move the decimal point three places to the right. Write in 0s for extra places. $45.62 \times 1{,}000 = 45620 = 45{,}620$

Use the quick way to multiply by 10, 100, or 1,000.

1) $.4 \times 10$

2) $.625 \times 10$

3) 1.8×10

4) 7.364×100

5) 89.2×100

6) $23.96 \times 1{,}000$

Lesson 37 Dividing Decimals

Words to Know

dividend the number to be divided
divisor the number to be divided by
quotient the answer in a division problem

Jeff bought three shirts for $74.85. How much did each shirt cost? You need to divide. Jeff spent $24.95 for each shirt.

A. Look at the three steps below to divide a decimal by a whole number.

Divide. 6.24 ÷ 5

Step 1 Find the decimal point in the **dividend.** Put the decimal point in the **quotient** above the decimal point in the dividend.

$5\overline{)6.24}$

Step 2 Divide just as you would divide whole numbers. Try to divide until there is no remainder. Add more 0s if you use all the digits of the dividend.

$$\begin{array}{r} 1.248 \\ 5\overline{)6.240} \end{array}$$

Sometimes you can divide a decimal and always get a remainder. You may decide to stop and round to 2 or 3 decimal places.

Divide to find the quotients.

1) $3\overline{)6.9}$ 2) $5\overline{)746.6}$ 3) $12\overline{)3.36}$

4) $54\overline{)7.128}$ 5) $18\overline{)168.48}$ 6) $64\overline{)58.56}$

B. To divide a decimal by a decimal, you must change the **divisor** to a whole number. Look at the four steps below.

Divide. 17.92 ÷ 6.4

Step 1 Make the divisor a whole number by moving the decimal point to the right of the last digit.

$64.\overline{)17.92}$

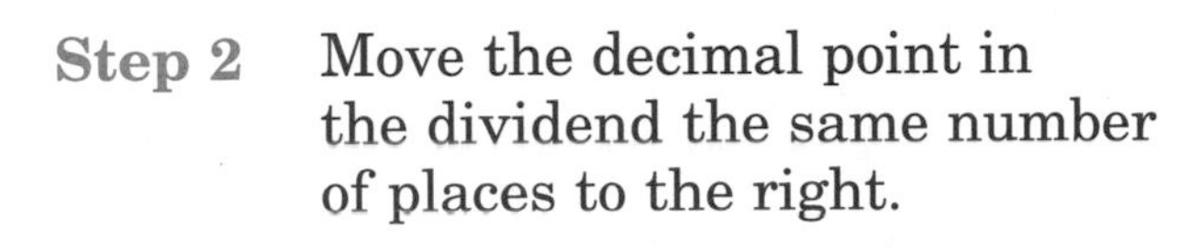

Step 2 Move the decimal point in the dividend the same number of places to the right.

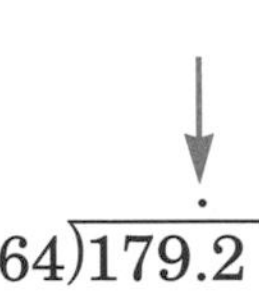

64)179.2

Step 3 Put a decimal point in the quotient above the decimal point in the dividend.

64)179.2

Step 4 Divide just as you would divide a decimal by a whole number.

$$\begin{array}{r} 2.8 \\ 64\overline{)179.2} \end{array}$$

Sometimes you need to add ending 0s to the dividend. Look at this example.

$.25\overline{)16.5}$ $\qquad$ $\begin{array}{r} 66 \\ 25.\overline{)1650.} \end{array}$

Divide to find the quotients.

1) $.3\overline{).6}$ 2) $.5\overline{)9.25}$ 3) $.79\overline{)4.661}$

4) $2.8\overline{).49}$ 5) $1.05\overline{)8.61}$ 6) $.08\overline{)57.6}$

Reminder

Show your work on a separate sheet of paper.

C. Dividing decimals by 10, 100, or 1,000 is not hard. Look at the examples below.

Divide. 45.2 ÷ 10 45.2 ÷ 1,000

Step 1 Count the zeros in the 10, 100, or 1,000.

Step 2 Move the decimal point one place to the left for each zero. Fill in with extra 0s if you need to.

There is **one** 0 in 10. Move the decimal point one place to the left. 45.2 ÷ 10 = 4.52 = 4.52

There are **three** 0s in 1,000. Move the decimal point three places to the left. Write in 0s for extra places. 45.2 ÷ 1,000 = .0452 = .0452

Use the quick way to divide by 10, 100, or 1,000.

1) 73.4 ÷ 10 2) 73.4 ÷ 100 3) 73.4 ÷ 1,000

4) .9 ÷ 10 5) 2.23 ÷ 1,000 6) 824.5 ÷ 100

Lesson 38 Changing Decimals to Fractions

Words to Know

fraction a number that names part of a whole

lowest terms when 1 is the only common factor of the numerator and the denominator

Sometimes you need to change a decimal to a fraction. Suppose you want to add .25, $\frac{1}{2}$, and $\frac{2}{3}$. First you must change .25 to a **fraction.**

A. To change a decimal to a fraction, you need to know the value of the decimal places. A place value chart can help. Look at the next two examples. Use the place value chart as you go.

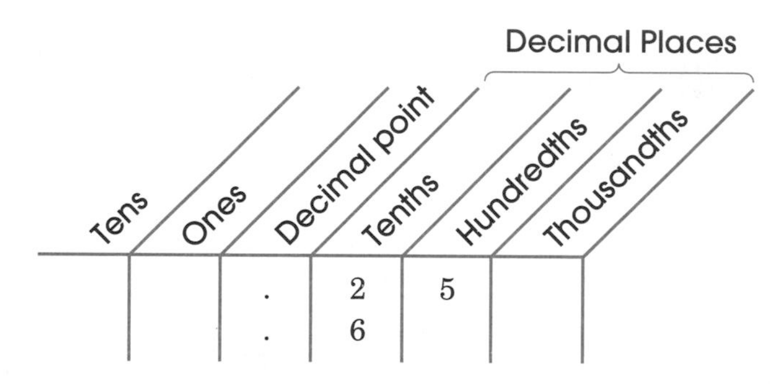

Change .25 to a fraction.

Step 1 Use the digits of the decimal as the numerator.
The numerator of .25 is 25.

$.25 = \frac{25}{}$

Step 2 Use the value of the decimal as the denominator.
The decimal .25 is 25 hundredths.

$.25 = \frac{25}{100}$

Step 3 Reduce the fraction to **lowest terms.**

$.25 = \frac{25}{100} = \frac{1}{4}$

$\frac{25}{100} = \frac{1}{4}$

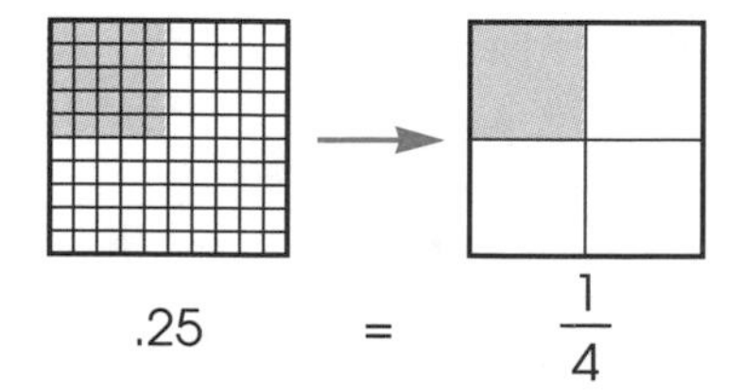

Here's another example.

$$.6 = 6 \text{ tenths} = \frac{6}{10} = \frac{3}{5}$$

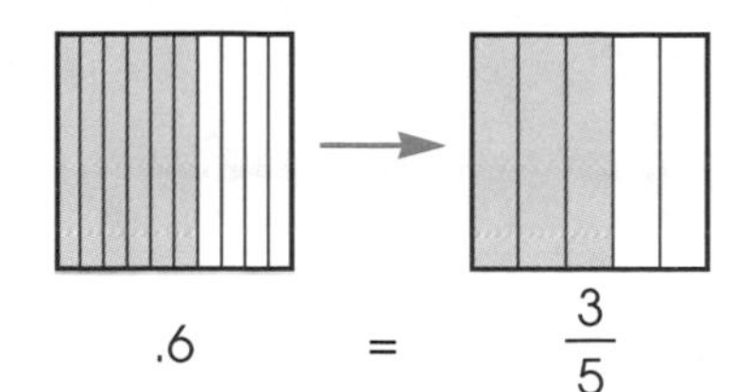

$.6 = \frac{3}{5}$

Change each decimal to a fraction. Reduce to lowest terms.

1) .5	2) .4	3) .9	4) .1
5) .05	6) .75	7) .8	8) .45

Reminder

To reduce a fraction to lowest terms, divide the numerator and denominator by their greatest common factor.

B.

If you have a mixed decimal change the decimal part to a fraction and keep the whole number on the left.

$$2.7 \longrightarrow 2 \text{ and } 7 \text{ tenths} \longrightarrow 2\frac{7}{10}$$

Change each mixed decimal to a mixed number.

1) 1.5	2) 4.8	3) 2.08	4) 5.75

C.

You also can change a fraction to a decimal. Divide the numerator by the denominator. Follow the example below.

Write $\frac{3}{4}$ as a decimal.

Step 1 Write the numbers as division. $3 \div 4$

Step 2 Divide 3 by 4. There are no 4s in 3. Add a decimal point and a zero to the 3. Put a decimal point in the quotient.

$$4\overline{)3.0}$$

Step 3 Divide as you would divide a decimal.

$$\begin{array}{r} .7 \\ 4\overline{)3.0} \\ -\ 28 \\ \hline 2 \end{array}$$

Step 4 There is a remainder.
Add another 0 to the dividend.
Bring down the 0.
Continue dividing until there is no remainder

$$\begin{array}{r} .75 \\ 4\overline{)3.00} \\ -\ 28 \\ \hline 20 \\ -\ 20 \\ \hline 0 \end{array}$$

$\frac{3}{4} = .75$

Write each fraction as a decimal.

1) $\frac{1}{2}$	2) $\frac{1}{4}$	3) $\frac{1}{5}$	4) $\frac{3}{6}$

Lesson 39 Rounding Decimals

Words to Know

rounding decimals changing decimals to say about how many wholes, tenths, or hundredths

A notebook costs $3.85. You can also say it costs about $4. When you use $4 instead of $3.85, you are rounding decimals.

A. **Rounding decimals** can help you talk about numbers. You can use a number line to round to the nearest whole number.

Round 3.2 to the nearest whole number.

- 3.2 is closer to 3 than to 4.
- 3.2 rounds down to 3.

Round 3.75 to the nearest whole number.

- 3.75 is closer to 4 than to 3.
- 3.75 rounds up to 4.

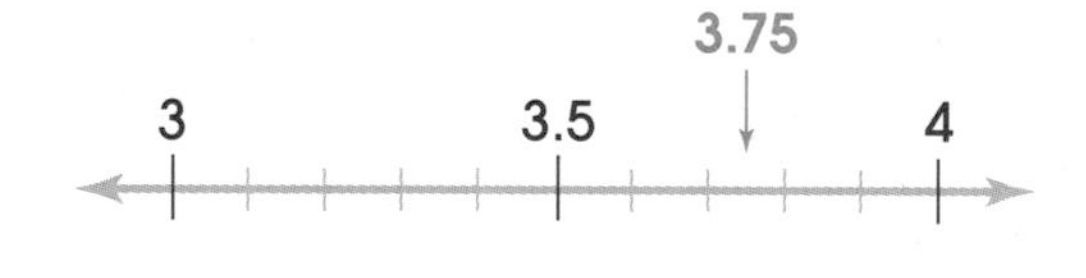

Round each decimal to the nearest whole number. Use the number line for help.

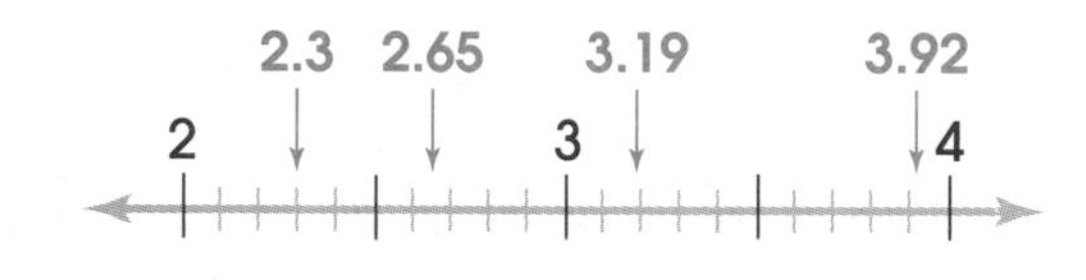

1) 2.3 2) 2.65 3) 3.19 4) 3.92

B. You do not need a number line to round decimals. You can round decimals almost the same way you round whole numbers.

Round 10.38 to the nearest whole number.

Step 1 Find the rounding place.
This problem says to round to the nearest *whole number.*

↓
10.38

Step 2 Look at the digit to the right.It is 3.
3 is less than 5. Make no changes to the digit in the rounding place.

↓
10.38

Step 3 Do not write anything to the right of the rounding place.

↓
10.38 rounds down to 10

Round each decimal to the nearest whole number.

1) 4.6 2) 12.1 3) 5.65 4) 23.52

C. You also can round to decimal place values. Follow the same steps.

Round 6.351 to the nearest *tenth.*

↓
6.351 rounds up to 6.4 because 5 equals 5.

Round 10.428 to the nearest *hundredth.*

↓
10.428 rounds up to 10.43 because 8 is greater than 5.

Round each decimal to the nearest *tenth.*

1) 5.58 2) .62 3) 9.57 4) 1.72

5) .22 6) 11.25 7) 14.09 8) 22.479

Round each decimal to the nearest *hundredth.* The first one is done for you.

9) 4.623 4.623 rounds down to 4.62

10) 8.019 11) 2.975 12) 9.004

13) 7.025 14) 6.229 15) 11.008

16) 28.961 17) 15.035 18) 14.602

Lesson 40 Decimals: Problem Solving

Words to Know

addition clue words words or groups of words that tell you to add

subtraction clue words words or groups of words that tell you to subtract

multiplication clue words words or groups of words that tell you to multiply

division clue words words or groups of words that tell you to divide

Clue words are important. They can help you decide how to solve a math problem. Read the clue words below.

addition clue words	in all, together, altogether, total, both
subtraction clue words	how many more, how many less, fewer, how much more, remain, left, difference, greater
multiplication clue words	*at* \$6 each, *for* 2 days, *in* 8 weeks
division clue words	into how many, how many times, how much/many did each

Addition, subtraction, multiplication, and division are called arithmetic operations. You can use these operations to solve problems with decimals. Some problems use more than one operation. Read the problem and the six steps below. Notice how the steps work.

Reminder

Not all problems have clue words.

Diane's mother gave her \$125 to buy a new outfit. Diane spent \$54.98 for a dress, \$32.95 for shoes, and \$15.98 for jewelery. How much did she have left from the \$125?

Step 1 Read the problem.

Step 2 Decide what you must find out. How much did Diane spend in all? How much money was left?

Step 3	Notice the clue words. Decide which operation(s)to use.	Left tells you to subtract. In all tells you to add. You will use two operations.	
Step 4	Write down the problems.	\$54.98 32.95 + 15.98 **total spent**	\$125.00 **– total spent** **money left**
Step 5	Solve the problems.	\$ 54.98 32.95 + 15.98 **\$103.91**	\$125.00 – 103.91 **\$21.09**
Step 6	Check the answer. Ask yourself if it makes sense.		

Read each problem below. Follow the steps to answer the questions. Then solve the problems.

1) Andy earned \$36.50 mowing lawns. He earned \$55.25 at a garage sale. He earned \$30 more helping a neighbor paint a fence. He needs \$200 to buy a set of weights. How much more does Andy need to earn?
   a. What are the clue words?
   b. Which operation(s) will you use?
   c. Solve the problem.
2) Linda bought four quarts of milk at \$.67 a quart. She bought five pounds of potatoes at \$.32 a pound. How much did she spend in all?
   a. What are the clue words?
   b. Which operation(s) will you use?
   c. Solve the problem.
3) Juan had \$235.67 in his checking account. He put in \$55. Then he wrote a check for \$35.98. How much did Juan have left in his account?
   a. What are the clue words?
   b. Which operation(s) will you use?
   c. Solve the problem.
4) Find the cost of 13.5 gallons of gasoline at \$1.42 a gallon.
   a. What are the clue words?
   b. Which operation(s) will you use?
   c. Solve the problem.
5) The Book Fair raised \$639.25. The money was shared by five clubs. How much money did each club receive?
   a. What are the clue words?
   b. Which operation(s) will you use?
   c. Solve the problem.

Unit 4 Percents

Lesson 41 What Is a Percent?

Word to Know

percent (%) a part of a whole that has been divided into 100 equal parts

Items in a store are often on sale. Suppose a jacket is 25 percent off. The number 25% is a percent. It means that for every $100 you spend, you will save $25.

The word *percent* sounds like per-SEHNT.

A. A **percent** is a part of a whole that has been divided into 100 equal parts. The symbol % is the sign for percent. Percent means "out of 100." For example, 25% means 25 out of 100. Look at the squares below. Notice that each big square is made of 100 little squares. A percent of each big square is shaded.

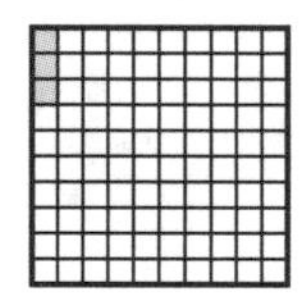

3%

3 out of 100 squares are shaded

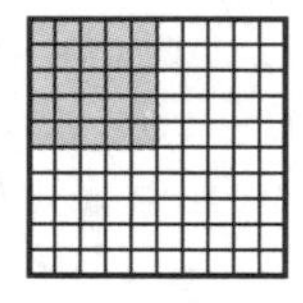

25%

25 out of 100 squares are shaded

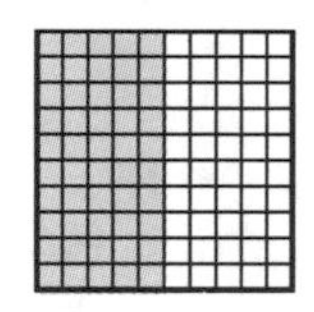

50%

50 out of 100 squares are shaded

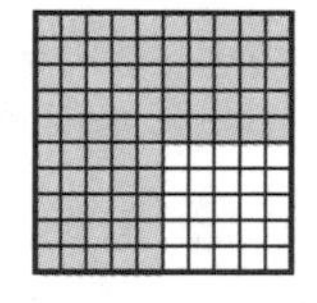

75%

75 out of 100 squares are shaded

100% means a whole or "all." The whole is shaded. All of the squares are shaded.

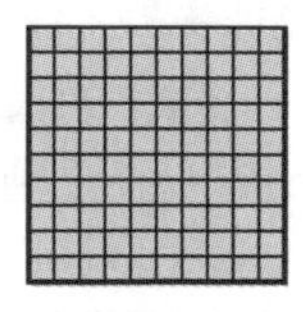

100%

100 out of 100 squares are shaded

Write the percent of each big square that is shaded. Give your answer in two ways.

1) 2) 3) 4)

B.

You can use percents to describe any part of 100.

There were 100 questions on a test.
Maria answered 80 of them.
→ Maria answered **80%** of the questions.

There are 100 cents in a dollar.
Tax is 5 cents for each dollar.
→ Tax is **5%**.

Write a sentence to answer the questions. Use a percent.

1) There are 100 students in the senior class. 28 of them walk to school. What percent walk to school?
2) There were 100 questions on a test. Sukie answered 93 of them correctly. What percent did she answer correctly?
3) 100 people were asked which soap they use. 67 said they use Kleen soap. What percent said they use Kleen?
4) There were 100 homework assignments last year. Craig did all of them. What percent of the homework assignments did Craig do?

C.

A percent can show more than a whole. Look at the picture below.

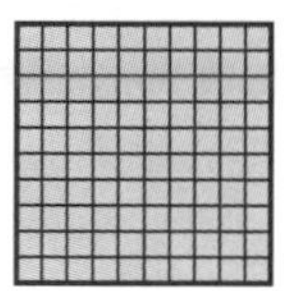

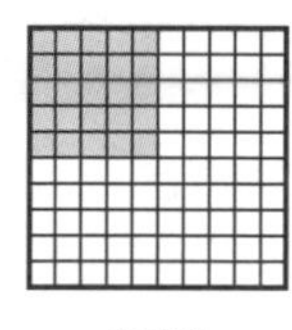

100% and 25%

One whole and 25% → 125%

Write the total percent shown by the shaded part of each pair of squares. Your answer will be *more* than 100%

1)

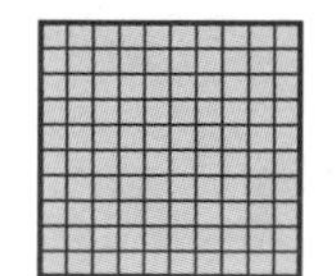

2)

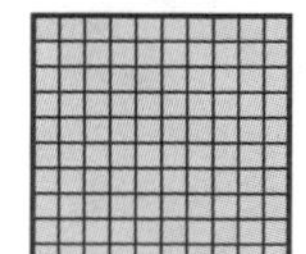

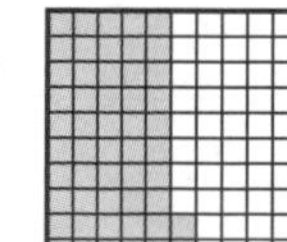

Lesson 42 Changing Percents to Decimals

Words to Know

decimal a number written with a dot followed by places to the right; the digits to the right of the dot stand for less than one whole

equivalent to be equal in value

The word *equivalent* sounds like ee-KWIHV-uh-lehnt.

There are different ways to show a number value. You can show it in a percent or a **decimal**. Remember a decimal is a number less than one whole. It is written with a dot followed by places to the right. Look at 50% and .50. They have **equivalent** values. This means they are equal in value.

50% = .50

A.

To multiply or divide a percent you must change the percent to an equivalent decimal. Changing percents to decimals is easy to do. Follow the three steps below.

Change 26% to a decimal.

Step 1	Take away the percent sign.	26
Step 2	If there is no decimal point, place a decimal point to the right of the last digit.	26.
Step 3	Move the decimal point two places to the left.	.26

When the percent is a single digit, write a 0 before moving the decimal point.

5% = .05 5 ⟶ 5. ⟶ .05 ⟶ .05

Change the percents to decimals.

1) 24%	2) 52%	3) 3%
4) 87%	5) 15%	6) 100%
7) 124%	8) 103%	9) 4%
10) 15.5%	11) 110%	12) 29%
13) 75%	14) 150%	15) 200%

B. Sometimes you will want to change a decimal to a percent. Look at the steps below.

Change .43 to a percent.

Step 1 Move the decimal point *two* places to the right. 43.

Step 2 Write a percent sign after the number. 43%

If the decimal is a whole number, add a decimal point after the last digit. Write two 0s before you move the decimal. Then write a percent sign.

5 = 500. = 500%

When you move the decimal, remember to add zeros if needed.

.3 = 30. = 30%

Change the decimals and whole numbers to percents.

1) .33	2) .4	3) .58
4) .69	5) .125	6) 1
7) .257	8) .82	9) .75
10) 1.2	11) .9	12) .07
13) .625	14) 2	15) 1.04

Lesson 43 Finding the Percent of a Number

Word to Know

percent (%) a part of a whole that has been divided into 100 equal parts

When you deal with money, you will often need to find the percent of a number. Suppose there is a 25%-off sale on sneakers. At full price, the sneakers cost $30. How much will you save? You need to find 25% of $30.

A. Finding the **percent** of a number is easy to do. Remember, percent is part of a whole divided into 100 equal parts. Follow the three steps below to find 25% of $30.

Find 25% of $30.

Step 1 Change the percent to a decimal. 25% = .25

Step 2 Multiply the other number in the problem by the decimal.

$$\begin{array}{r} \$30 \\ \times\ .25 \\ \hline 150 \\ 600 \\ \hline \$7.50 \end{array}$$

Step 3 Put the decimal point in the correct place in the product.

25% of $30 is $7.50. To check, remember 25% is less than a whole. So 25% of 30 should be less than 30.

Reminder

To change a percent to a decimal, move the decimal point 2 places to the left.

Find the percents of the numbers given.

1) 10% of 90

2) 20% of 34

3) 8% of 80

4) 25% of 42

5) 60% of 100

6) 60% of 50

7) 60% of 200

8) 12% of 85

B. There is a quicker way to find 10% of a number. Look at the two examples below. Notice in the answers the digits are the same. The decimal point is moved 1 place to the left.

Longer Way	Quicker Way
Find 10% of 75. 10% → .10 → .1 $\begin{array}{r} 75 \\ \times\ .1 \\ \hline 7.5 \end{array}$	Find 10% of 75 7.5 **7.5**

Longer Way	Quicker Way
Find 10% of 342 10% → .10 → .1 $\begin{array}{r} 342 \\ \times\ .1 \\ \hline 34.2 \end{array}$	Find 10% of 342 34.2 **34.2**

Here's another way to think about it.

Find 10% of 90.
Move the decimal point 1 place to the left. 9 0 → 9.0
Drop the zero at the end of the decimal. 9

Find 10% of each number. The first one is done for you.

1) 80 *10% of 80 is 8.*
2) 93 3) 20 4) 100 5) 125

C. Knowing how to find percents can help you every day.

Solve the problems below.

1) A scout troop is trying to collect 1,000 cans. So far, the scouts have collected 50% of their goal. How many cans have they collected?
2) Jason saw a basketball he wanted. Yesterday it was \$22. Today it is on sale for 20% off. If he buys the basketball on sale, how much will he save?
3) The gym has 400 seats. The basketball game was 33% sold out. How many seats were sold?
4) The concert was 10% sold out. The arena has 1,250 seats. How many were sold?

Lesson 44 Percents: Problem Solving

Words to Know

addition clue words words or groups of words that tell you to add

subtraction clue words words or groups of words that tell you to subtract

multiplication clue words words or groups of words that tell you to multiply

We may use percents when we want to solve a problem. Clue words can be useful. They can help us decide which math operation to use. Read the clue words below.

addition clue words	20% more
subtraction clue words	20% off, how many are left?, how much more?
multiplication clue words	20% of

You can solve a word problem with percents. Read the word problem below. Then follow the five steps.

> Ted's Hobby Shop is having a big sale. All model rockets are 25% off the regular price. The usual price of the rocket you want is $18. How much will it cost on sale?

Step 1 Read the problem. Look for clue words. — 25% off

Step 2 Decide what you must find out.
(1) How much is 25% of $18?
(2) How much will the rocket cost after you subtract the savings?

Step 3	Find 25% of $18.	18 × .25 90 360 **$4.50**
Step 4	Subtract that amount from $18.	**$18.00** − 4.50 $13.50
Step 5	Think about your answer. Decide if it makes sense.	Yes; on sale the rocket costs less.

Solve the problems below. Use the steps you learned above.

1) William makes $5 per hour working at a restaurant. He will soon receive a 5% raise. Then how much will he make per hour?
2) Kathryn bought a dress at a 20%-off sale. The price of the dress had been $27. How much did Kathryn pay?
3) A class is trying to raise $500 to go on a field trip. The students have raised 40% of $500. How much more do they need to raise?
4) Jim bought an old bike for $10. He fixed it up and sold it for 200% of what he paid for it. How much did he sell it for?
5) There are 16 students in a math class. On the last test, 75% of them received an A or B. How many students received an A or B?
6) The high school is presenting a spring play. There are 600 seats in the theater. The play is 62% sold out. How many seats are left to sell?
7) Tom went to a hardware store sale. Everything in the store was 20% off. He also had a coupon for 10% off the sale price of any item. He bought tools that regularly cost $20. How much did he pay *after both* of his savings? (*Hint:* take off the 20% first. Then take off the 10%.)
8) How much does a $8 toy cost with 5% tax?

Glossary

addition putting numbers together; finding the *total* amount *page 8*

addition clue words words or groups of words that tell you to add: in all, together, all together, total, both, more *pages 38, 64, 80, 88*

arithmetic operations addition, subtraction, multiplication, and division *page 38*

common denominator a common multiple of two denominators *page 46*

common factors factors that two products have that are the same *page 35*

compare to look at two different things and see how they are the same or different *page 36*

comparing decimals finding which decimal is larger or smaller: 99.21 is larger than 99.12 *page 68*

comparing fractions looking at fractions to decide which is larger or smaller: $\frac{3}{4}$ is larger than $\frac{2}{4}$; $\frac{1}{2}$ is larger than $\frac{1}{3}$ *page 48*

cross cancellation dividing a numerator and a denominator by the same number *page 56*

decimal a number written with a dot followed by places to the right; the digits to the right of the dot stand for less than one whole: .2, .05, .468, .1372 *pages 66, 84*

decimal places the places to the right of the decimal point—tenths, hundredths, thousandths: 34.568 has 3 decimal places *page 72*

decimal point the dot in a decimal number: .50 *pages 66, 70*

denominator the bottom number in a fraction; the denominator tells how many parts are in the whole: in the fraction $\frac{5}{8}$, the denominator is 8; there are 8 parts in the whole *page 40*

difference the amount you get when one number is subtracted from another: the difference of 9 and 3 is 6 *page 12*

digit used to write a number: 0, 1, 2, 3, 4, 5, 6, 7, 8, 9 *page 4*

divide to find out how many times a number contains another: $10 \div 2 = 5$ *page 26*

dividend the number to be divided: the dividend in $10 \div 2$ is 10 *pages 28, 74*

division dividing *page 26*

division clue words words or groups of words that tell you to divide: in, how many, into how many, how many times, how many did each *pages 38, 64, 80*

divisor the number to divide by: the divisor in $10 \div 2$ is 2 *pages 28, 74*

equal (=) the same as *page 8*

equivalent to be equal in value: 25% = .25 *page 84*

equivalent fractions fractions that name the same amount: $\frac{1}{2}$ and $\frac{2}{4}$ *page 42*

even number a number that ends in 0, 2, 4, 6, or 8: 72 and 458 are even numbers *page 3*

factor one of the numbers multiplied to find a product: 2 and 3 are factors of 6 *pages 18, 34*

fraction a number that names part of a whole: $\frac{1}{2}$, $\frac{3}{4}$ *pages 40, 76*

greatest common factor (GCF) the largest factor that two or more products have that are the same: 3 is the greatest common factor of 6 and 9 *page 35*

improper fraction a fraction whose numerator is larger than its denominator: $\frac{5}{3}$, $\frac{10}{7}$ *pages 50, 52*

invert to reverse, or switch, the positions of the numerator and denominator in a fraction: $\frac{3}{4}$ inverted is $\frac{4}{3}$ *page 58*

least common denominator the smallest common denominator *page 47*

least common multiple (LCM) the smallest multiple that two or more numbers share that is not 0: 12 is the least common multiple of 3 and 4 *page 36*

like fractions fractions that have the same denominator: $\frac{2}{3}$ and $\frac{1}{3}$, $\frac{3}{7}$ and $\frac{6}{7}$ *pages 46, 60*

lowest terms fraction with 1 as the only common factor of the numerator and denominator: $\frac{2}{3}$ is in lowest terms *page 45*

mixed decimal a number containing a whole number and a decimal number: 4.5, 36.13, 128.559 *page 67*

mixed number a number that is made up of a whole number and a fraction: $3\frac{2}{5}$, $12\frac{3}{4}$ *pages 50, 52*

multiple the product of multiplying a number by other whole numbers: 0, 4, 8, and 12 are multiples of 4 *page 36*

multiplication multiplying two or more numbers *page 18*

multiplication clue words words or groups of words that tell you to multiply: **of, part,** at $4 each, **for** 3 days, **in** 6 weeks, **per** hour, 20% **of** *pages 38, 64, 80, 88*

multiply to add a number one or more times: the addition problem 3 + 3 + 3 + 3 = 12 is the same as the problem 4 × 3 = 12 *page 18*

number line numbers shown in order as points on a line *page 3*

numerator the top number in a fraction; the numerator tells how many parts are being used: in the fraction $\frac{2}{3}$, the numerator is 2 *page 40*

odd number a number that ends in 1, 3, 5, 7, or 9: 25 and 137 are odd numbers *page 3*

partial product the number you get when you multiply a number by one digit of another number: In 24 × 12, 48 and 240 are the partial products *page 22*

percent (%) a part of a whole that has been divided into 100 equal parts *pages 82, 86*

place value value or amount of a digit; a digit's worth: the digit 7 in 175 is in the tens place; it has a place value of 70 *page 4*

product the answer to a multiplication problem: the product of 3 and 4 is 12 *pages 18, 24, 54*

proper fraction a fraction whose numerator is smaller than its denominator: $\frac{3}{4}$, $\frac{6}{8}$, $\frac{8}{11}$ *page 50*

quotient the answer in a division problem: the quotient of 10 ÷ 2 is 5 *pages 26, 28, 30, 32*

reduce to divide the numerator and the denominator by the same number: $\frac{2}{3}$ is in lowest terms *pages 44, 55*

regroup to rename a number in order to add, subtract, multiply, or divide *pages 10, 14, 16, 20, 23*

remainder (R) the number left over in a division problem: 11 ÷ 2 = 5 R1 *pages 30, 32*

rename to write a number in a different way: 12 can be renamed as 1 ten + 2 ones *pages 5, 10, 14, 16*

rounding decimals changing decimals to say *about* how many in wholes, tenths, or hundredths *page 78*

rounding numbers changing numbers to say about how many tens, hundreds, or thousands: 58 rounded to the nearest ten is 60; 216 rounded to the nearest hundred is 200 *page 6*

subtraction clue words words or groups of words that tell you to subtract: how many more, how many fewer, how much more, how many less, remain, left, difference, greater, fewer, 20% off *pages 38, 64, 80, 88*

subtraction taking away one number from another; finding the difference *page 12*

sum the amount you get when numbers are added; the *total:* the sum of 3 and 5 is 8 *page 8*

total the answer to an addition problem *page 70*

unlike fractions fractions that have different denominators: $\frac{1}{2}$ and $\frac{1}{3}$, $\frac{3}{5}$ and $\frac{7}{8}$ *pages 46, 62*

whole numbers 0, 1, 2, 3, 4, 5, 6, 7, and so on *page 2*